STUDENT UNIT GUIDE

AS Chemistry

UNIT 1

AQA

Module 1: Atomic Structure, Bonding and Periodicity

Neil Goldie

Philip Allan Updates
Market Place
Deddington
Oxfordshire
OX15 0SE

tel: 01869 338652
fax: 01869 337590
e-mail: sales@philipallan.co.uk
www.philipallan.co.uk

ISBN 0 86003 725 8

This Guide has been written specifically to support students preparing for the AQA AS Chemistry Unit 1 examination. The content has been neither approved nor endorsed by AQA and remains the sole responsibility of the author.

Typeset by Good Imprint, East Grinstead
Printed by Information Press, Eynsham, Oxford

P00011

Contents

Introduction

About this guide

This guide is for students following the AQA Chemistry AS specification. It deals with Unit 1, which examines the content of **Module 1: Atomic Structure, Bonding and Periodicity**. This unit covers 30% of the total AS marks, or 15% of the total A-level marks. The $1\frac{1}{2}$ hour exam consists of several structured questions, of varying length, which are all compulsory. Some questions require short answers while some are more extended, needing answers written in continuous prose. There are 90 marks available on this paper.

Module 1 is probably the most important, because success in all future modules depends on an understanding of the basic principles and concepts covered here.

The key to success

It is essential that you can recall the basic facts and definitions, but a deeper understanding of the subject is essential if you want to achieve the top grades. The key to success in chemistry is *to understand the fundamental concepts* and *be able to apply them to new and unfamiliar situations*. Good examination technique is also an important factor, enabling you to work more effectively in the exam and gain access to the marks needed for a top grade.

This guide allows you to look again at the content of the specification, to test yourself at the end of each section and to assess your own work. It is essential that you read through the examiner's comments because they will help you improve your exam technique. Once you have completely worked through this guide, you will be very aware of your weak areas and it is these areas that you really need to address. Make a list of these weak areas and discuss any problems with other students in your class and with your teacher.

Using this guide

This guide has three sections:

- **Introduction** — this provides guidance on study and revision, together with advice on approaches and techniques to ensure you answer the examination questions in the best way that you can.
- **Content Guidance** — this section is not intended to be a textbook. It offers guidelines on the main features of the content of Unit 1, together with particular advice on making study more productive.
- **Questions and Answers** — this shows you the sort of questions you can expect in the unit test. Grade-A answers are provided; these are followed by examiner's

comments. Careful consideration of these will improve your answers and, much more importantly, will improve your understanding of the chemistry involved.

Revise a topic using the Content Guidance section as a guide. If there is something you do not understand, you should also refer to your own class notes and textbooks.

It is essential that you write down *specific* questions and discuss them with your teacher. For instance, 'Please could you explain bonding again' will not lead to good use of your time if you understand most of the features of ionic, covalent and metallic bonds. 'Please could you explain what is meant by the term coordinate bond and give me some examples' is more specific and shows that you have worked hard to identify weak areas.

Once you have revised a particular topic thoroughly, you should attempt the relevant question in the Question and Answer section, *without* looking at the grade-A answer.

Compare your answer with the grade-A answer and estimate your own performance. A rough guide to use is 80% = grade A, 70% = grade B, 60% = grade C, etc. However, these grade boundaries are adjusted, depending on the performance of the candidates.

Read through the examiner's comments to find out if you have made any of the common mistakes and to see how you could improve your technique. The comments also give some alternative answers.

Make a note of *specific questions* that caused you problems and discuss them with other students and your teacher.

Revision schedule

- Plan your revision schedule carefully.
- It is essential that you revise regularly.
- Leave yourself enough time to cover all the material. You need to go through each topic once as a basic minimum and then go through the weak areas again.
- In the weeks leading up to the exam, it is the weak areas that you should be revising, not every topic.
- In each revision session do not try to achieve too much. Revise one topic per session, e.g. ionisation energies. Here is one way to structure your session:
  - revise from the Content Guidance section (and your own notes)
  - make a brief written summary (no more than an A4 sheet of paper)
  - attempt the question
  - mark your answer
  - read the examiner's comments

- If you have scored a grade A, then tick the topic in the table shown below. You must have a break before you start the revision of the next topic.
- If there are weak areas and questions that you clearly do not understand, then write down specific questions ready for discussion with your teacher.
- Finally, make sure that you attempt all the past paper questions from the exam board and study the mark schemes carefully. Questions may be repeated or similar questions set.

Revision checklist

Once you have completed your revision and you feel that you understand the topic fully, tick the relevant box. If you have identified a weak area, place a cross in the relevant box.

Atomic structure

Topics	Details	✓ or ✗
Structure of atoms	Protons, neutrons, electrons Mass number and atomic number Isotopes	
Mass spectrometry	Basic principles of operation Interpretation of spectra, to give A_r Relative molecular mass	
Electron arrangement	*s*, *p*, *d* levels for atoms and ions up to atomic number 36	
Ionisation energies	Definition Variation across period 3 Variation down group 2 Successive ionisation energies	

Amount of substance

Topics	Details	✓ or ✗
Relative A_r and M_r	Definitions	
Mole and the Avogadro constant	Calculations using $n = m/A_r$ Calculations	
Ideal gas equation	Recall the equation Use it to calculate M_r	
Empirical and molecular formulae	Definitions Calculation of empirical formula Deduction of molecular formula	
Equations	Deduction of molecular formula Writing equations for reactions Reacting masses Reacting volumes Titration calculations	

Bonding

Topics	Details	✓ or ✗
Types of bonding	Ionic Covalent Dative covalent (coordinate) Metallic Bond polarity Polarisation of ions	
Intermolecular forces	van der Waals forces Dipole–dipole forces Hydrogen bonding	
States of matter	Solids, liquids and gases	
Types of crystal	Ionic Molecular Macromolecular Metallic	
Shapes of molecules	Basic principles and examples	

Periodicity

Topics	Details	✓ or ✗
Classifying elements	*s*, *p*, *d* block elements	
Physical properties of period 3 elements	Atomic radius Melting and boiling points First ionisation energies Electronegativity Conductivity	
Physical properties of group 2 elements	Atomic radius First ionisation energies Electronegativity Melting points	
Group 2 chemistry	Reactions of elements with water Solubility of hydroxides Solubility of sulphates Test for a sulphate	
Atypical nature of beryllium	Covalent nature of $BeCl_2$ Amphoteric nature of $Be(OH)_2$ Four coordination in aqua-complexes	

Unit Test 1

If you have revised thoroughly, completed all the questions in this guide and discussed problems with other students and your teacher, you should enjoy the exam. If you have completed all the AQA past papers, then the style of the paper will be very familiar and you will recognise some questions in the exam because they will be similar to previous questions.

Do not begin to write as soon as you open the paper — quickly scan the questions first.

It is *not* essential that you answer the questions in order. If the first question is difficult, then leave it to the end. It *is* essential that you answer *all* the questions.

You will have enough time to answer all the questions, provided that you keep your answers concise and you do not include irrelevant information. It is easy to waste time writing out a section of your notes that is irrelevant to the question asked. Do not repeat the question when starting your answer. The key to exam success is achieving the maximum number of marks in the minimum number of words.

The mark allocation at the end of each question should be used to estimate the amount of detail needed in your answer. If there is 1 mark available, the examiner is looking for a key word or phrase and certainly no more than one sentence. If there are 4 marks available, you should include four key points, which usually means writing four short sentences.

No marks are available for producing neat answers, but it certainly helps the examiners when they are marking your work. Untidy diagrams may become inaccurate and this definitely loses marks.

Terms used in examination questions

Give/state/name

These appear in many of the early structured questions. You need only write one or two words.There is usually 1 mark available for questions of this type.

Define

It is essential that you learn all your definitions. You need to state a definition or law very concisely. Definitions are usually worth 2 or 3 marks and the mark allocation indicates how many key points must be included.

Complete

You have to finish off a diagram, a graph or a table.

Draw/show by means of a diagram

Most drawings require only a simple sketch. It must be clear and accurate, but the examiner does not expect it to be of the same standard as that seen in your textbook. Poor presentation can make diagrams inaccurate and this is when marks are lost.

Calculate or determine

Use information to calculate a final answer that must be shown to the correct number of significant figures. Always show your working and include appropriate units after the final answer.

Write a balanced equation

In chemical equations, state symbols are not usually expected unless the examiner specifically asks for them. In physical processes and thermochemistry equations, state symbols are usually expected. For example, the first ionisation energy of sodium is:

$Na(g) \longrightarrow Na^+(g) + e^-$

Describe/explain

'Describe' means only give a description, whereas 'explain' requires a reason or interpretation. Both these terms mean more depth is required in your answer. You can judge how much detail is required by the mark allocation. 10 marks means that ten key points must be covered and this will probably require you to write a minimum of ten sentences. In general, most candidates write too much and give far too much detail.

Use

Questions using this term often include some data. Make sure you do use the data and include them in your answer. You may have to use a basic principle and apply it to a new and unfamiliar situation.

Suggest

This means that you will probably not have come across the material in the question before. You will have been taught the basic principle and now you have to apply this knowledge to an unfamiliar situation.

Content Guidance

This section covers the content of **Module 1: Atomic Structure, Bonding and Periodicity**. It is the most important module at AS and A-level because it covers many basic concepts that must be fully understood to guarantee success in all units.

In this Content Guidance section, the specification has been converted into user-friendly language and is in a format that is easy to remember. All the key facts, definitions and basic principles are covered. In order to achieve a top grade, it is essential that you fully understand the basic concepts and that you can apply them to unfamiliar situations.The content of this module falls into four sections, which are summarised below.

Atomic structure

Much of this has been covered at GCSE. The new topics are ionisation energies and the mass spectrometer, which are regularly tested. These new topics appear quite demanding at the beginning of the course. However, questions on these topics are usually well answered. Electron arrangements are now in terms of sub-levels.

Amount of substance

This difficult topic is usually covered in the early stages of the AS course. Many students find the concepts difficult to grasp and the topic difficult to revise. The only way to revise calculations effectively is to practise lots of different examples from past papers and textbooks. Repeating the process makes you more confident and you develop your own method for answering each type of question. Sometimes the calculation will still be too difficult, but questions on this topic also require recall of some basic definitions, giving you access to some of the marks.

Bonding

This builds on many of the basic concepts covered at GCSE. However, dative covalency, electronegativity, polarisation and different types of intermolecular force are all new concepts. This section can be quite demanding because of all the new ideas, so it is essential to simplify the material. 'Shapes of molecules' is a new topic which most candidates find easy, if taught in a systematic way.

Periodicity

This is a good topic to test your understanding of Atomic structure and Bonding. If you understand the basic concepts of structure and bonding, it is easy to explain the observed trends in the physical properties of the elements of period 3 and group 2. A new and more demanding topic is the atypical properties of beryllium.

Atomic structure

Fundamental particles — protons, neutrons and electrons

This first topic builds on the basic ideas covered at GCSE. Questions on this topic are relatively straightforward and are usually worth only a few marks in the early structured questions.

The properties of the **fundamental particles** are summarised in the table below.

Particle	Relative charge	Relative mass
Proton	+1	1
Neutron	0	1
Electron	–1	1/1840

The relative mass of an electron is so small that *negligible* or *zero* are also acceptable answers.

An atom consists of **electrons** (e) surrounding a nucleus that contains **protons** (p) and **neutrons** (n).

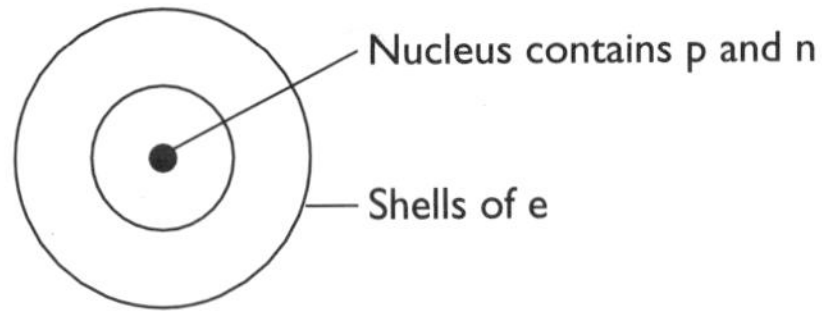

Mass number and atomic number

The symbols of the elements give information on the **atomic number** and the **mass number**, which in turn reveal the number of fundamental particles in each atom of the element.

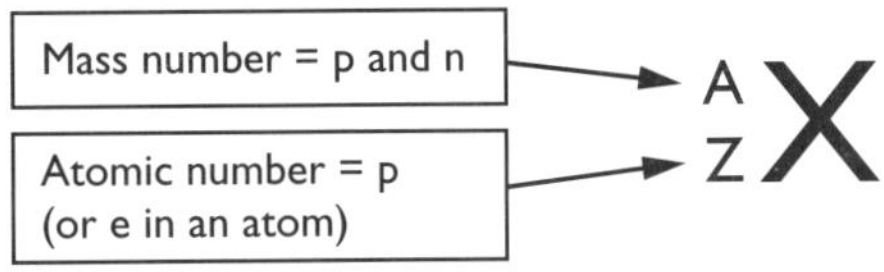

The number of protons and neutrons never changes in ion formation, but the number of electrons does.

$^{A}_{Z}X^{2-}$

This negative ion has two additional electrons

$^{A}_{Z}X^{3+}$

This positive ion has lost three electrons

Example

State the number of fundamental particles in the following species.

Species	Protons	Electrons	Neutrons
$^{31}_{15}P$	15	15	16
$^{24}_{12}Mg^{2+}$	12	10	12
$^{35}_{17}Cl^{-}$	17	18	18

Isotopes

The isotopes of magnesium have been used to illustrate all the basic principles, but other elements could be used. The three isotopes of magnesium are shown below, with the number of protons, electrons and neutrons listed for each isotope.

$^{24}_{12}Mg$	$^{25}_{12}Mg$	$^{26}_{12}Mg$
12p	12p	12p
12e	12e	12e
12n	13n	14n

- Isotopes have the same number of protons and same number of electrons, but a *different number of neutrons.*
- Isotopes are the same element with the *same atomic number* but a *different mass number.*

The mass spectrometer

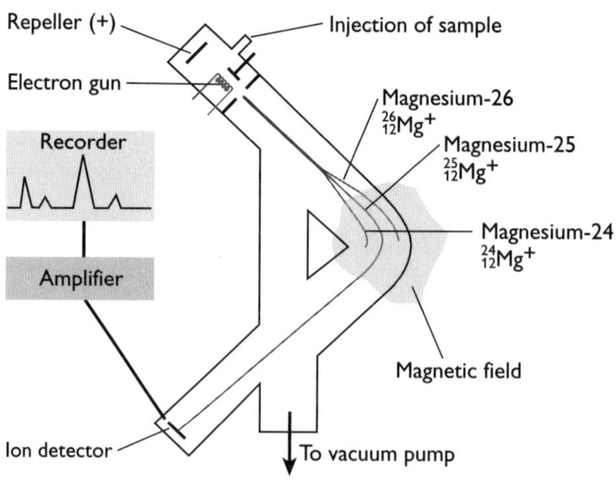

The principles of a simple mass spectrometer are limited to ionisation, acceleration, deflection and detection.

The apparatus illustrated can be used to separate the three isotopes of magnesium. If a diagram is required in the unit test, then a simple sketch would be sufficient. If a diagram is provided, you should be able to label it, particularly the paths of the different ions. In the same strength of magnetic field, the ion with the smallest mass-to-charge ratio is deflected the most.

An explanation of how the mass spectrometer is used to separate isotopes is a common question. The most common mistake is the inclusion of too much detail. There are four key stages in the separation of the isotopes and you need to mention four key points for each stage. The four stages are ionisation, acceleration, deflection and detection.

Ionisation

- Vaporise the sample.
- Bombard it with electrons using an electron gun.
- This knocks out another electron.
- Positive ions (cations) are produced.

The equation $Mg(g) + e^- \longrightarrow Mg^+(g) + 2e^-$ could score 4 marks.

Acceleration

- The cations are accelerated.
- Passing through an electric field achieves this.
- Ions move towards a negative plate.
- Ions are focused into a beam by slits.

Deflection

- The ions are deflected.
- Passing through a magnetic field achieves this.
- Deflection depends on the mass-to-charge ratio (m/z).
- The smaller the m/z ratio, the greater is the deflection.

Detection

- Ions hit a detector.
- Current is produced.
- Information is fed to a computer or displayed on a chart.
- Varying either the electric field or magnetic field collects ions of different masses.

Interpretation of simple mass spectra of elements

The mass spectrum of magnesium is shown overleaf.

- It has three peaks confirming that three isotopes are present.
- The peak at m/z value 24 has the highest relative intensity, so this means it is the most abundant isotope.

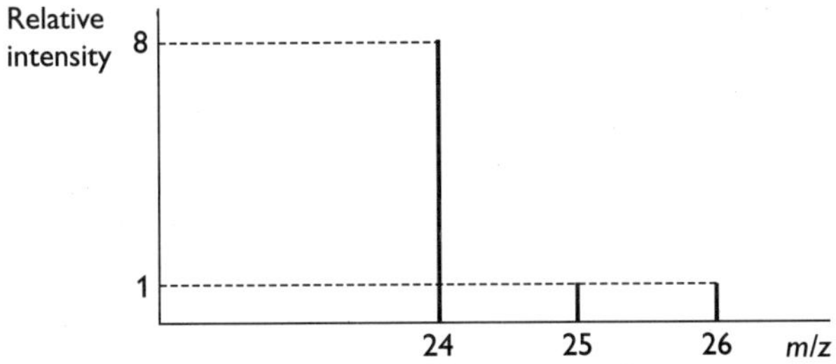

- Peaks can be observed at m/z values of 12, 12.5 and 13 in the mass spectra of magnesium. Some ions with a +2 charge can be produced when two electrons are knocked out. These ions have half the m/z ratio of the +1 ions.

Calculation of relative atomic mass from isotopic abundance

This is limited to mononuclear ions.

Example 1: the relative atomic mass of magnesium

Using the mass spectrum of magnesium, it is possible to determine the average relative atomic mass (A_r).

$$A_r = \frac{(8 \times 24) + (1 \times 25) + (1 \times 26)}{10} = 24.3$$

Data are presented in different ways and careless mistakes are often made in the calculations.

Example 2: the relative atomic mass of boron

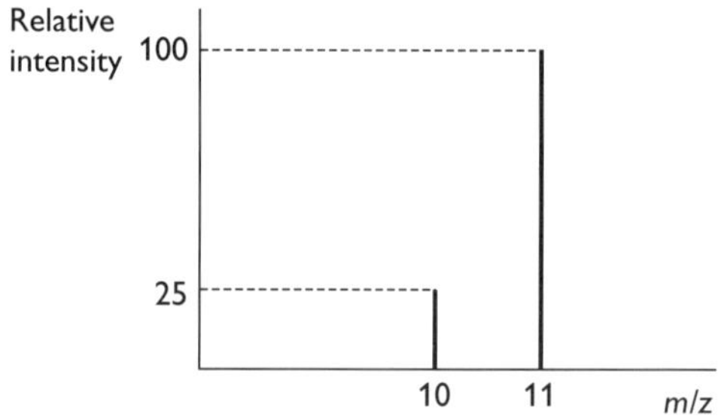

Common mistake

$$\frac{(25 \times 10) + (100 \times 11)}{100} = 13.5 \text{ ✗✗✗}$$

Correct answer

$$\frac{(25 \times 10) + (100 \times 11)}{125} = 10.8 \text{ ✓✓✓}$$

Example 3: identifying an unknown element

m/z	80	82	83	84	86
Relative intensity	1	5	5	25	8

$$A_r = \frac{(80 \times 1) + (82 \times 5) + (83 \times 5) + (84 \times 25) + (86 \times 8)}{\text{total intensity of 44}} = 83.9$$

The answer is **krypton** — a common mistake is to give the answer as polonium. Remember, when identifying an element using the mass spectra data, you must look at *mass* numbers and not *atomic* numbers.

The use of mass spectrometry to determine relative molecular mass

The spectra of compounds can be really complicated when they break up and produce smaller ions. The M_r of the compound can be determined from spectra by looking at the peak with the largest m/z value.

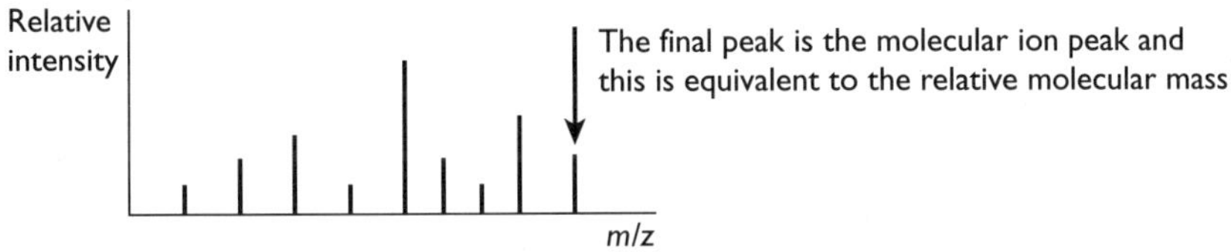

Electron arrangements

Very few questions are set on electron arrangements. However, it is essential that the basic principles are understood because they are needed to explain many other topics. Consider the energy diagram below for the first 36 electrons.

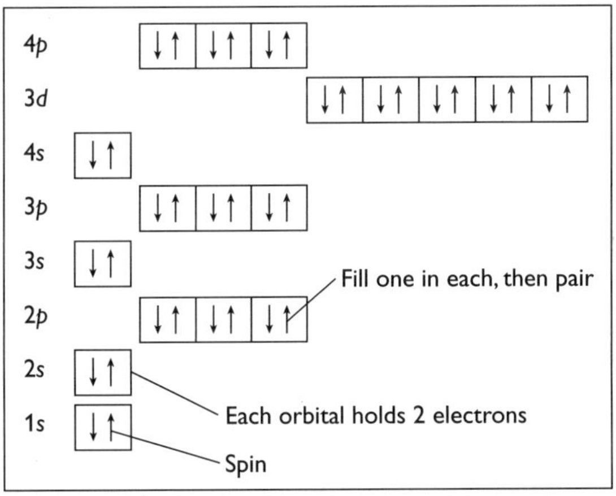

Refer to this diagram when reading through the following key points.

- Electrons are arranged in energy **shells** from shell 1 (lowest energy) to shell 4 (highest energy).
- Each shell contains **sub-shells** designated *s*, *p*, *d* and *f*.
- Shell 1 contains one sub-shell (*s*), shell 2 contains two sub-shells (*s* and *p*), shell 3 contains three sub-shells (*s*, *p* and *d*), shell 4 contains four sub-shells (*s*, *p*, *d* and *f*).
- Each sub-shell contains a different number of **orbitals** and each orbital can hold a maximum of two electrons.
- An ***s* sub-shell** contains one orbital and so can hold two electrons; a ***p* sub-shell** contains three orbitals and so can hold six electrons; and a ***d* sub-shell** contains five orbitals and so can hold ten electrons.

- When filling up the energy diagram above, remember that *electrons enter the lowest energy levels first.*
- When filling a 2*p* sub-shell, three electrons are first placed in separate orbitals [↑] [↑] [↑] so that they are **unpaired**. Then the next three electrons are added to complete the sub-shell [↑↓] [↑↓] [↑↓]. This arrangement is also followed when filling the 3*d* sub-shell.
- The main complication is the electron arrangements of the transition metal atoms and their ions. Looking at the energy diagram above it is clear that the 4*s* level is filled before the 3*d*. However, *when forming ions, the 4s is lost before the 3d.*
- Two transition elements have unusual electron configurations. These are chromium and copper:
  Cr $1s^2\,2s^2\,2p^6\,3s^2\,3p^6\,4d^5\,4s^1$
  Cu $1s^2\,2s^2\,2p^6\,3s^2\,3p^6\,3d^{10}\,4s^1$

The examples below show the electron configurations of some common atoms and ions. The number of electrons in each sub-shell is always written as a superscript, for example $1s^2$, not 1*s*2.

$_{15}P$	$1s^2\,2s^2\,2p^6\,3s^2\,3p^3$
$_{19}K$	$1s^2\,2s^2\,2p^6\,3s^2\,3p^6\,4s^1$
$_{30}Zn$	$1s^2\,2s^2\,2p^6\,3s^2\,3p^6\,4s^2\,3d^{10}$ (usually written $1s^2\,2s^2\,2p^6\,3s^2\,3p^6\,3d^{10}\,4s^2$)
$_{26}Fe$	$1s^2\,2s^2\,2p^6\,3s^2\,3p^6\,3d^6\,4s^2$
$_{26}Fe^{2+}$	$1s^2\,2s^2\,2p^6\,3s^2\,3p^6\,3d^6$
$_{26}Fe^{3+}$	$1s^2\,2s^2\,2p^6\,3s^2\,3p^6\,3d^5$

Ionisation energies

The definition of the first ionisation energy **X(g) ⟶ X⁺(g) + e⁻** covers three key points. It is the energy required to convert **one mole** of **gaseous atoms** into one mole of **gaseous cations**, each atom losing an electron.

Variation across the periodic table

The graphs presented in many textbooks appear far too complicated. Keep your graph simple and remember the pattern '2–3–3, then big drop'.

This 2–3–3 pattern is the same for both period 2 and period 3.

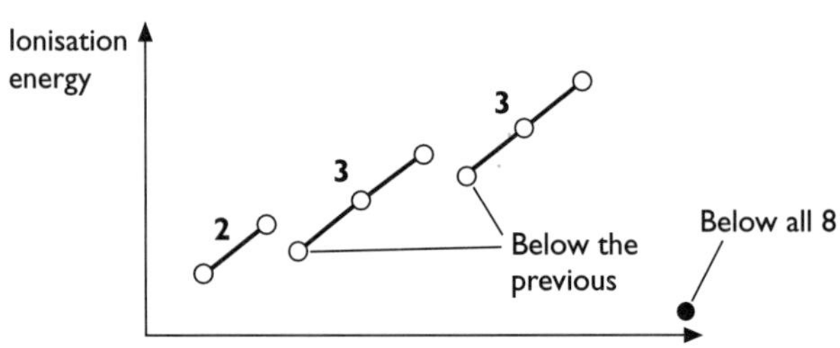

Now join the points up to complete the graph. This shows the variation across period 2 or period 3.

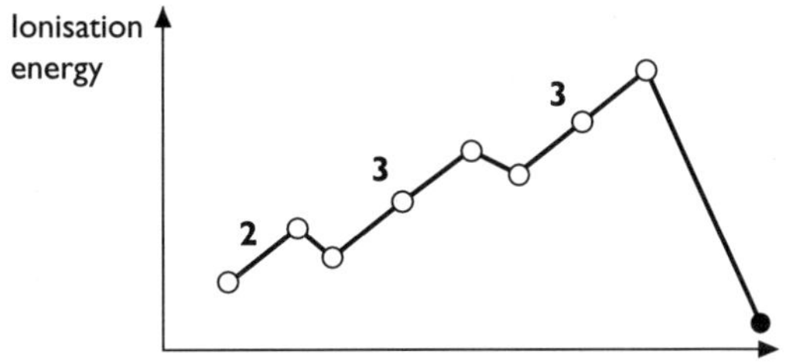

You could be asked to explain the following features of the graph:

- the general increase across the period
- the drops between groups 2 and 3 and between groups 5 and 6
- the big drop at the end of a period

These are very common questions.

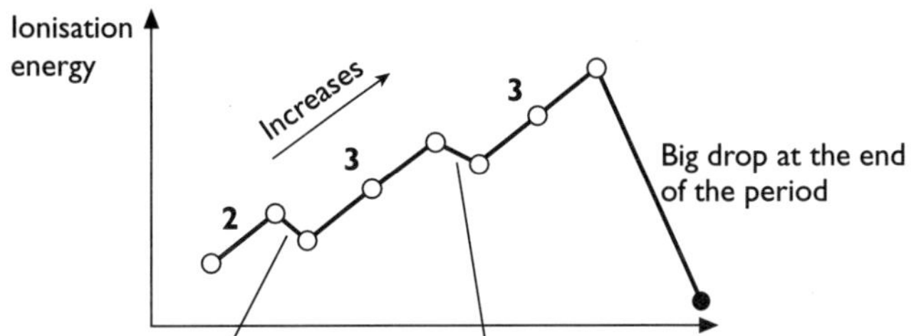

General increase across the period

- Increase in proton number (or nuclear charge).
- Shielding the same (or electrons enter the same shell).
- Increased attraction for electrons.

Drop between groups 2 and 3

Drawing the electron arrangements makes it much easier to score the marks:

Mg: $1s^2\,2s^2\,2p^6\,3s^2$

Al: $1s^2\,2s^2\,2p^6\,3s^2\,3p^1$

- The electron is removed from the **3*p* level**.
- 3*p* is higher in energy than 3*s*, so the electron is easier to remove.

The same argument can be used for beryllium and boron, but the electron is removed from the 2*p* level, which is higher in energy than the 2*s*.

Drop between groups 5 and 6

Again, draw the electron arrangements of the two atoms and emphasise the arrangement of the electrons in the 3*p* level.

P: $1s^2\,2s^2\,2p^6\,3s^2\,3p^3$ ↑ ↑ ↑

S: $1s^2\,2s^2\,2p^6\,3s^2\,3p^4$ ↑↓ ↑ ↑

- There are **paired electrons** in one of the 3*p* orbitals in sulphur.
- This causes **repulsion**, so the electron is easier to remove.

The same argument can be used for nitrogen and oxygen, but the electrons are paired in one of the 2*p* orbitals.

The big drop at the end of a period

This is the same explanation as for the decrease in ionisation energies down any group.

- More shells (or increase in atomic radii).
- More shielding.
- Less attraction for outer electrons.

Successive ionisation energies

Second ionisation energy:

$X^{+}(g) \longrightarrow X^{2+}(g) + e^{-}$

Third ionisation energy:

$X^{2+}(g) \longrightarrow X^{3+}(g) + e^{-}$

Successive ionisation energies allow us to deduce how many electrons are in the outside shell of an element.

- Successive ionisation energies increase — the same number of protons and fewer electrons mean more attraction for the remaining electrons.
- Outer electrons are easier to remove because they are shielded by the inner electrons.
- The big jump in ionisation energy indicates the electron is removed from a new inner shell, which is closer to the nucleus and less shielded by other electrons.

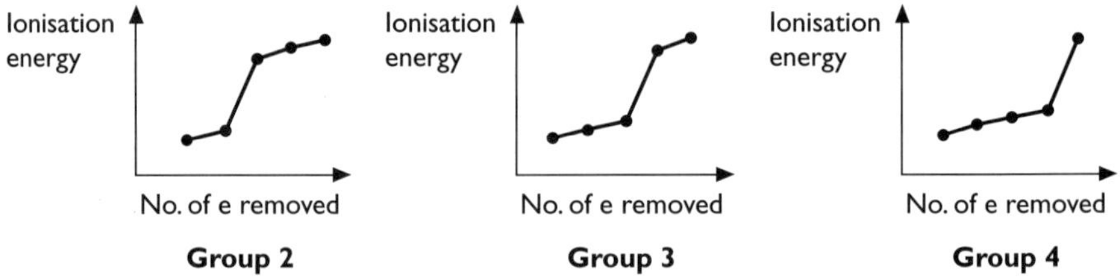

Amount of substance

This is probably the most difficult section in Module 1. If you find calculations difficult, then take comfort in the fact that many candidates are in the same position. You can spend a long time revising this section and still find that the question in the exam is too difficult.

The questions involving calculations are usually very structured, in order to guide you through to the final answer. Many consequential marks are available. This means that if your answer to question (a) is wrong, making your answers to (b), (c) and (d) also incorrect, you can still score full marks for questions (b) to (d), provided that your reasoning is correct and all your working has been clearly shown.

I have outlined the basic definitions or principles that you need to know in each topic and given example calculations where appropriate. You must *practise as many calculations as possible* from books, past papers and from your class notes.

Relative masses, the mole and the Avogadro constant

Definitions

- The **relative atomic mass** is defined as the mass of one atom of an element relative to $\frac{1}{12}$ of the mass of one atom of carbon-12, where carbon is 12.0000:

$$A_r = \frac{\textbf{mass of one atom of an element}}{\frac{1}{12}\textbf{ mass of one atom of }^{12}\textbf{C}}$$

- The **relative molecular mass** of a compound is defined as the mass of one molecule of the compound relative to $\frac{1}{12}$ of the mass of one atom of carbon-12.

$$M_r = \frac{\textbf{mass of one molecule of the compound}}{\frac{1}{12}\textbf{ mass of one atom of }^{12}\textbf{C}}$$

- The **Avogadro constant** is the number of particles in one mole of any substance. This constant number of particles is **6.023×10^{23} particles mol^{-1}** and is given the symbol **L**.
- The **mole** is defined as the amount of substance that contains as many elementary units (e.g. atom, molecule, ion, electron, etc.) as there are atoms in 12 g of carbon-12. It is given the unit **mol**.
- The **molar mass** is the mass of one mole of an element or a compound. For example, NaOH is $40\,g\,mol^{-1}$.
- The number of moles is related to the mass and the M_r of the substance by the following equation:

$$\textbf{number of moles } (n) = \frac{\textbf{mass}}{\textbf{molar mass}} = \frac{m}{M_r}$$

It is the last equation that you need to be able to use and rearrange.

Examples

- **A sodium atom has a mass of 3.817×10^{-23} g. Calculate the total mass of 1 mol of sodium atoms.**

The mass of 1 mol = mass of 1 atom × Avogadro constant
= $3.817 \times 10^{-23} \times 6.023 \times 10^{23} = 22.99$ g

- **Calculate the mass of 0.5 moles of carbon dioxide (CO_2).**

Mass = number of moles × M_r
= $0.5 \times 44 = 22$ g

- **Calculate the number of moles in 10 kg of sulphuric acid (H_2SO_4).**

$$\text{Number of moles} = \frac{\text{mass}}{\text{molar mass}}$$

$$= \frac{10 \times 10^3}{98} \quad \text{(because 10 kg is } 10 \times 10^3 \text{ g)}$$

$$= 102$$

Molarity

- A **molar solution** is made by dissolving a mole of a substance in sufficient water to make 1 dm^3 (or 1 litre) of solution. This gives a solution of concentration 1 mol dm^{-3}. For example, a molar solution of sodium hydroxide is made by dissolving 40 g of solid sodium hydroxide in water and then making the solution up to a total volume of 1 dm^3.
- A solution of known concentration is known as a **standard solution**. Standard solutions are made up in **volumetric flasks** of varying volumes.
- The relationship between number of moles, mass and molarity can be summarised as follows:

$$\text{number of moles} = \frac{\text{mass}}{\text{molar mass}} = \frac{\text{molarity} \times \text{volume}}{1000}$$

$$n = \frac{m}{M} = \frac{MV}{1000}$$

Examples

- **What mass of aminosulphonic acid (H_2NSO_3H) will be dissolved in 250 cm^3 to produce a solution of 0.100 mol dm^{-3} concentration?**

Number of moles = $MV/1000 = 0.1 \times 250/1000 = 0.025$ mol

Mass = number of moles $\times M_r = 0.025 \times 97 = 2.425$ g = 2.43 g (3 sig. fig.)

- **What is the concentration, in mol dm^{-3}, of a solution containing 4 g of NaOH in 200 cm^3 of solution?**

Number of moles = mass/M_r = 4/40 = 0.1 mol

Molarity = number of moles $\times 1000/V = 0.1 \times 1000/200 = 0.5$ mol dm^{-3}

Empirical and molecular formulae

Definitions

- The **empirical formula** is the *simplest whole number ratio* of the *atoms* in a compound. The empirical formula of hydrogen peroxide is HO.
- The **molecular formula** is the *actual number of atoms* present in the compound. The molecular formula is a simple multiple of the empirical formula. The molecular formula of hydrogen peroxide is H_2O_2.

Examples

- **A chloride of iron contains 65.5% chlorine. Calculate the empirical formula.**

	Iron	Chlorine
$\frac{m}{M}$	$\frac{34.5}{56.0}$	$\frac{65.5}{35.5}$
Moles	0.616	1.85
Ratio	1	3
Formula	$FeCl_3$	

- **A compound has the empirical formula CH_2O and the molar mass of $180\,g\,mol^{-1}$. Deduce the molecular formula.**

$CH_2O = 30$
$M_r = 180$
$180/30 = 6$
Molecular formula = 6 × empirical formula = $C_6H_{12}O_6$

Balanced equations and associated calculations

Reacting masses from balanced equations

The masses of the products and the reactants involved in a reaction can be deduced from the stoichiometry (the number of moles of each substance) of the balanced equation.

Consider the thermal decomposition of potassium hydrogencarbonate:

$$
\begin{array}{lclclcl}
2KHCO_3 & \longrightarrow & K_2CO_3 & + & CO_2 & + & H_2O \\
2\,mol & \longrightarrow & 1\,mol & + & 1\,mol & + & 1\,mol \\
2(100)\,g & \longrightarrow & 138\,g & + & 44\,g & + & 18\,g \\
200\,g & \longrightarrow & 138\,g & + & 44\,g & + & 18\,g
\end{array}
$$

Using the data above, it is possible to predict the amount of each substance used or produced in the reaction.

Examples

- **What will be the mass of the solid residue when 10 g of potassium hydrogencarbonate are heated?**

$$
\begin{array}{lclclcl}
2KHCO_3(s) & \longrightarrow & K_2CO_3(s) & + & CO_2(g) & + & H_2O(g) \\
200\,g & \longrightarrow & 138\,g & + & 44\,g & + & 18\,g \\
10\,g & \longrightarrow & 6.9\,g & + & 2.2\,g & + & 0.9\,g
\end{array}
$$

The amount of solid residue remaining is 6.9 g

- **What mass of quicklime (CaO) can be obtained by heating 75.0 g of limestone, which is 90.0% calcium carbonate?**

Amount of calcium carbonate present in the limestone is 75 × 90/100 = 67.5 g

$CaCO_3$	$\longrightarrow$	CaO	+	CO_2
100 g	$\longrightarrow$	56 g	+	44 g
67.5 g	$\longrightarrow$	37.8 g	+	29.7 g

The mass of quicklime obtained is 37.8 g

Many questions involving reacting masses are set on an industrial scale. This means that kilograms or tonnes may be used. Do not be intimidated by this type of question, because if 1 g of substance X gives 2 g of substance Y then 1 kg of X gives 2 kg of Y. The questions are basically the same, but with different units quoted in your answer.

Calculating molarities and volumes for reactions in solution

The type of calculations in the exam will vary, but the most straightforward ones can be divided into three main stages.

- When you know both the molarity (M) and the volume (V), use the formula $\boldsymbol{n = MV/1000}$ to calculate the number of moles of solution.
- Write the equation for the reaction (or use the equation that has been given) and use the ratio of the reactants in the equation to predict the number of moles of the other solution.
- You know the values of n (from stage 2) and V (different from the one used in stage 1), so rearranging $n = MV/1000$ allows you to determine the value of M. Always quote your answer to the correct number of significant figures.

Examples

- **A 25.0 cm³ portion of 0.100 mol dm⁻³ sodium hydroxide required 22.5 cm³ of ethanoic acid for complete reaction. Calculate the concentration of the ethanoic acid.**

The number of moles of sodium hydroxide = $MV/1000 = 0.1 \times 25/1000 = 2.5 \times 10^{-3}$

Using the equation, the number of moles of ethanoic acid can be deduced (1:1 ratio).

$$CH_3COOH + NaOH \longrightarrow CH_3COONa + H_2O$$

2.5×10^{-3} $\qquad\qquad$ 2.5×10^{-3}

The concentration of ethanoic acid in mol dm^{-3} is found using $n = MV/1000$ again.

$$2.5 \times 10^{-3} = M \times 22.5/1000$$

Rearranging gives $M \times \dfrac{2.5 \times 10^{-3} \times 1000}{25} = 0.110\ \text{mol dm}^{-3}$

- **What volume of 0.185 mol dm⁻³ sodium hydroxide solution will react completely with 20.0 cm³ of 0.100 mol dm⁻³ sulphuric acid?**

The number of moles of sulphuric acid = $MV/1000 = 0.1 \times 20/1000 = 2 \times 10^{-3}$

Using the equation, the number of moles of sodium hydroxide can be deduced (1:2 ratio):

$$H_2SO_4 + 2NaOH \longrightarrow Na_2SO_4 + 2H_2O$$
$$2 \times 10^{-3} \quad 4 \times 10^{-3}$$

The concentration of sodium hydroxide in mol dm^{-3} is found using $MV/1000$ again.

$n = MV/1000$

Rearranging gives $V = n \times 1000/M = 4 \times 10^{-3} \times 1000/0.185 = 21.6\,cm^3$

The ideal gas equation

You need to be able to recall the ideal gas equation and then use it to calculate relative molecular masses or to predict, from a given equation, the amount of gas produced. The most common mistake is made when converting the units given in the question. If you cannot attempt the calculation, then quote the equation pV = nRT because it is usually worth 1 mark.

Symbol	Name	Units	Notes
p	Pressure	Pa or kPa	Atmospheric pressure is 101 kPa, which is 1.01×10^5 Pa
V	Volume	m^3	$1\,m^3 = 1 \times 10^3\,dm^3 = 1 \times 10^6\,cm^3$ $1\,cm^3 = 1 \times 10^{-3}\,dm^3 = 1 \times 10^{-6}\,m^3$
n	Moles	Mol	
R	Gas constant	$J\,K^{-1}\,mol^{-1}$	This value remains constant at 8.31
T	Temperature	K	0° C = 273 K (if temperature is in °C, then add 273)

Calculating the relative molecular mass (M_r) of a gas

$$pV = nRT \quad \text{where } n = \frac{\text{mass}}{\text{relative molecular mass}} = \frac{m}{M_r}$$

$$pV = \frac{mRT}{M_r}$$

$$\text{So } M_r = \frac{mRT}{PV}$$

Examples

- **Calculate the M_r of krypton. 1.809 g of krypton has a volume of 500 cm^3 at a temperature of 0 °C and a pressure of 9.8×10^4 Pa.**

Using the above equation, this calculation be shown in one step.

$$M_r = \frac{1.809 \times 8.31 \times 273}{9.8 \times 10^4 \times 500 \times 10^{-6}} = 83.75$$

- **What volume of oxygen at 25 °C and 101 kPa is evolved from the decomposition of 1.7 g of hydrogen peroxide?**

  $$2H_2O_2(l) \longrightarrow 2H_2O(l) + O_2(g)$$

Moles of H_2O_2 = m/M_r = 1.7/34 = 0.05 mol

Using the 2:1 ratio in the equation, the moles of oxygen = 0.025 mol

Rearranging $pV = nRT$ gives

$$V = nRT/p$$
$$= 0.025 \times 8.31 \times 298/101 \times 10^3$$
$$= 6.13 \times 10^{-4}\,\text{m}^3 = 0.613\,\text{dm}^3$$

Bonding

The nature of ionic, covalent and metallic bonds

Try to remember the characteristic features of each type of bonding as a list and then convert the list into the sentences you need for the exam. Your answers should be concise and only include relevant material. You must be able to recognise the type of bonding when presented with different elements or compounds.

Ionic bonding

The main points are:

- transfer of electrons
- transfer is from metals to non-metals
- metals lose e^- to form cations (+ ions)
- non-metals accept e^- to form anions (– ions)
- oppositely charged ions are held together by strong electrostatic forces of attraction
- the forces of attraction result in a **giant,** three-dimensional, ionic, crystalline **lattice**
- examples include CsF, MgO, NaCl, KBr

The ionic lattice of sodium chloride is illustrated below.

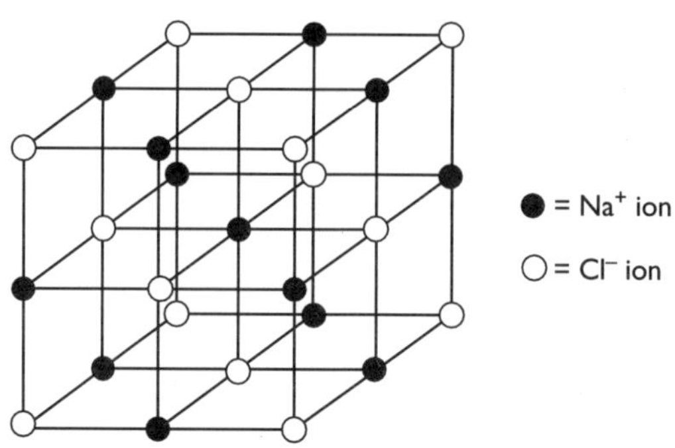

Covalent bonding

The main points are:

- electrons are shared
- sharing is between two non-metals
- non-metal atoms both need to gain electrons — this is achieved by sharing
- a covalent bond is defined as a **shared pair of electrons**
- covalent bonds can exist in isolation in simple molecules
- examples include Cl_2, F_2, CH_4, NH_3, H_2O, HCl, HF

A simple molecule of hydrogen fluoride is shown below.

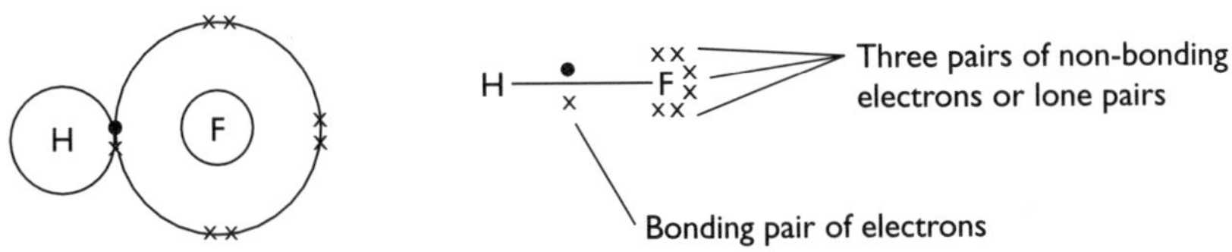

Coordinate or dative covalent bonding

The main points are:

- a covalent bond is defined as a shared pair of electrons
- a coordinate bond or a dative covalent bond is defined as a *shared pair of electrons* where *both electrons have originated from one atom*
- once a coordinate bond has formed, it is identical to a covalent bond because it is a shared electron pair

Example 1: ammonia and boron trifluoride

Ammonia gas and boron trifluoride gas react readily to give a white solid with the composition NH_3BF_3.

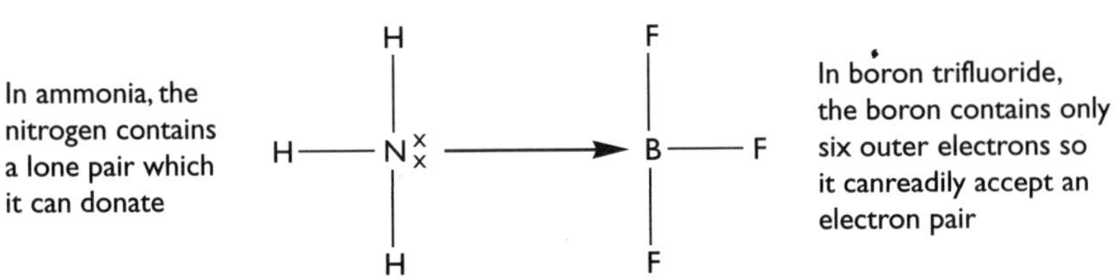

Example 2: ammonia and the H^+ ion

$H_3N\!: \longrightarrow H^+$ produces NH_4^+ (the ammonium ion).

Metallic bonding

The main points are:

- A metal is a giant three-dimensional lattice, consisting of closely packed metal ions surrounded by delocalised electrons.
- Delocalised electrons are free to move through the lattice.

- A metallic bond is the electrostatic force of attraction that two neighbouring cations have for the delocalised electrons between them.
- Examples include any metal, such as sodium, potassium, calcium, iron.

A lattice of sodium metal is illustrated below.

positive ions should not touch!

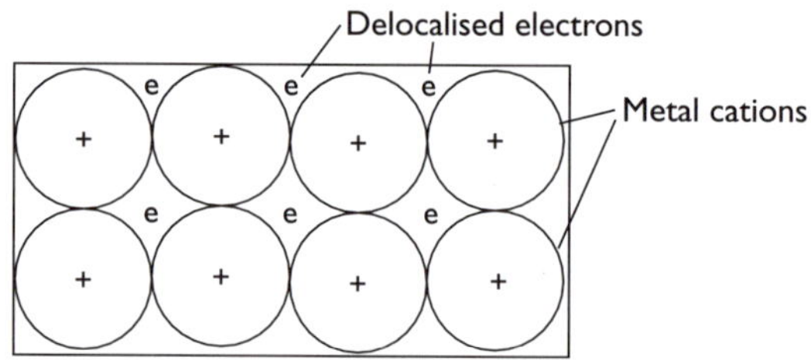

Bond polarity and the polarisation of ions

Bond polarity

You will encounter questions on bond polarity if the **covalent molecule** contains *atoms with differing electronegativities*. You need to understand the concept of electronegativity.

An electron pair shared equally between two atoms constitutes a covalent bond. Sometimes, this electron pair is not shared equally because some atoms attract electrons more than others. The unequal sharing of electrons is known as **bond polarity**. It represents an introduction of some ionic character into the covalent bond. **Electronegativity** is a measure of the *relative tendency of an atom to attract a bonding pair of electrons* (or to withdraw electron density) *from a covalent bond*. Elements have electronegativity values ranging from 0 to 4. You need to remember the trends, not the actual numbers.

- Electronegativity *increases across a period*.
- Electronegativity *decreases down a group*.
- Fluorine is the most electronegative element in the periodic table.

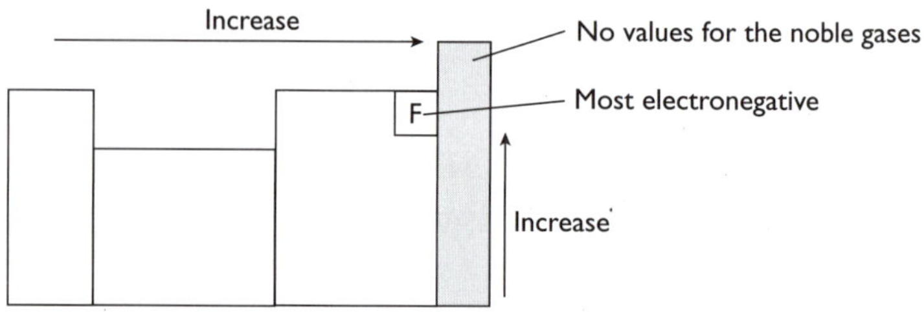

- Small atoms with a large number of protons in the nucleus attract electron density more strongly.
- No values are quoted for the noble gases because these elements generally do not form compounds with covalent bonds.

When a covalent bond exists between atoms of differing electronegativity, the shared pair of electrons is displaced towards the more electronegative atom. The charge separation creates an electric dipole and the molecule is described as polar. The greater the difference in electronegativity between the elements, the more polar is the compound.

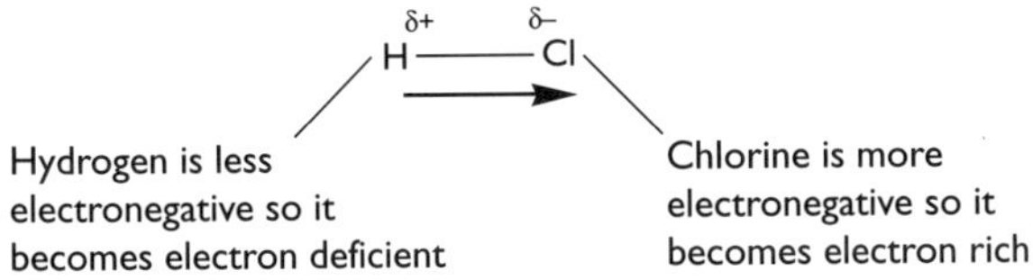

In **symmetrical molecules**, such as carbon dioxide, the bond polarities cancel, so there is *no overall polarity*.

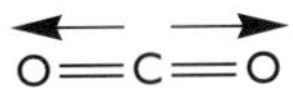

Polarisation of ions

You will encounter questions on polarisation if the ionic compound contains a small cation and a large anion (e.g. lithium and beryllium compounds).

An ionic compound: caesium fluoride

Ionic compounds are formed when electron transfer occurs between a metal and a non-metal to produce oppositely charged ions (a positive cation and a negative anion), which attract each other electrostatically to produce giant, three-dimensional, ionic lattices.

When the ions are *perfectly spherical*, then the bonds are *perfectly ionic*.

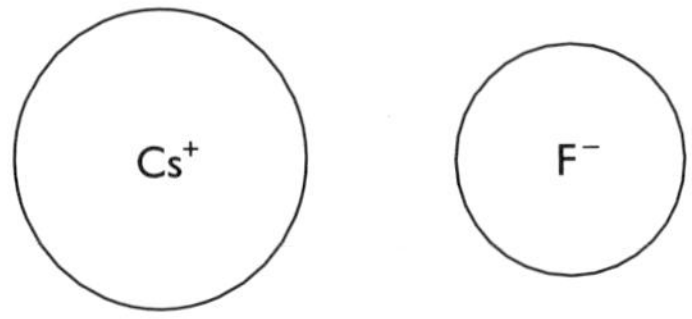

An ionic compound with covalent character: lithium iodide

When the cation is small or has a high charge, it has a high charge density. A small, highly charged cation is very polarising. A polarising cation withdraws electron density from the anion and distorts the electron cloud of the anion. The anion is no longer spherical, but is distorted, and the ionic compound now has covalent character. In lithium iodide, the lithium is the small polarising cation and the iodide is the large polarisable anion.

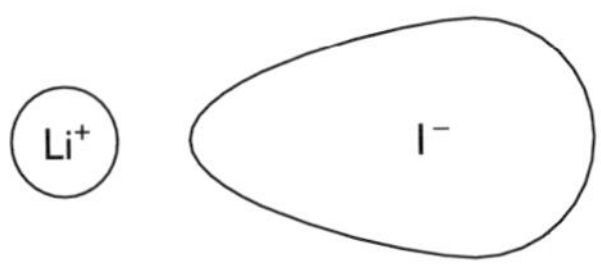

Predicting the extent of covalent character

- Covalent character is shown in ionic compounds if the cation is polarising.
- The greater the charge density of the cation (i.e. small size and high charge), the greater is its polarising power.
- The anions are more polarisable as they increase in size.

Ionic compounds	Extent of covalent character
$BeCl_2$ and $MgCl_2$	$BeCl_2$ has more covalent character because the Be^{2+} ion is smaller than Mg^{2+} and so it is more polarising
NaCl and $MgCl_2$	$MgCl_2$ has more covalent character because the Mg^{2+} ion is smaller and has a higher charge than Na^+ and so it is more polarising
LiI and LiF	LiI has more covalent character because the I^- ion is larger and more polarisable than the F^- ion and so it is more distorted

Forces acting between molecules

You need to understand the different types of intermolecular force so that you can explain the physical properties of substances, such as boiling points and solubility.

Covalent molecules held together by intermolecular forces

There are three types of intermolecular force. In order of increasing strength they are:

- van der Waals forces
- permanent dipole–dipole
- hydrogen bonding

Predicting the type of intermolecular force

van der Waals forces attract all covalent molecules to each other.

Polar molecules contain atoms with different electronegativities, so in addition to van der Waals forces, the molecules also attract each other by permanent dipole–dipole forces.

If a molecule contains hydrogen, which is directly bonded to either a nitrogen, oxygen or fluorine atom, it sets up an extreme form of a dipole–dipole force between the molecules. This is hydrogen bonding.

An example of van der Waals forces: iodine

- Iodine (I_2) is a non-polar molecule.
- The iodine atoms have identical electronegativities.
- Electron distribution is usually symmetrical.

- Temporary dipole occurs when electron distribution is asymmetric (unequal) due to fluctuating movement of electrons.
- Iodine molecules are large. The van der Waals forces between the molecules are therefore relatively strong compared with other, simple, non-polar molecules.
- Remember, compared with other intermolecular forces, the van der Waals forces in iodine are still weak.

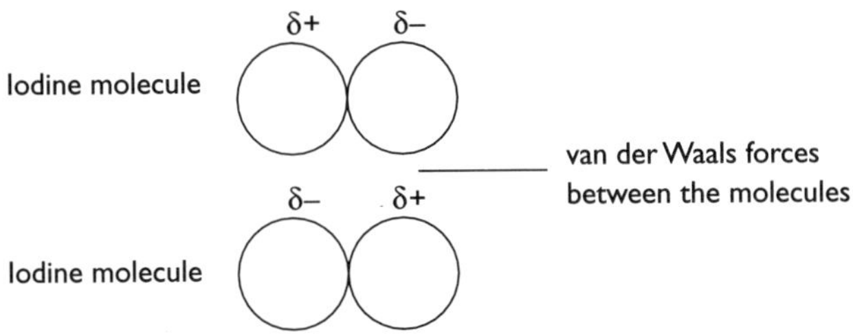

Other examples include chlorine, bromine, methane, phosphorus and sulphur.

An example of dipole-dipole forces: hydrogen chloride

- Hydrogen chloride (HCl) is a polar molecule.
- The hydrogen and chlorine atoms have different electronegativities.
- Chlorine has a greater electronegativity than hydrogen.
- Electron distribution is asymmetrical (i.e. permanent charge separation).
- Permanent dipole–dipole forces exist between the molecules.
- These forces are stronger than van der Waals forces, when comparing molecules of similar size.

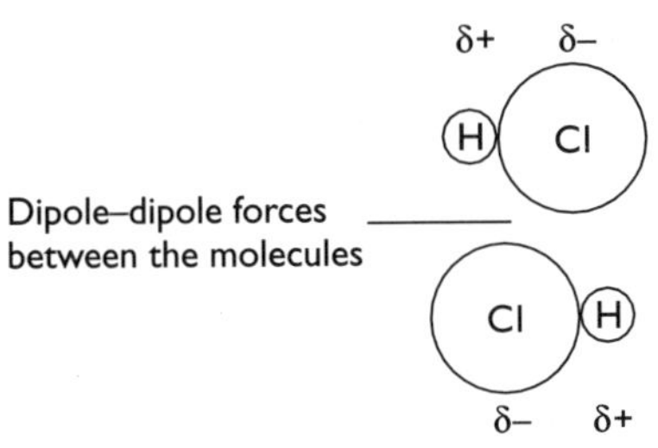

Other examples include hydrogen bromide, hydrogen iodide, chloromethane and hydrogen sulphide.

An example of hydrogen bonding: water

- Water (H_2O) is a polar molecule.
- The hydrogen is directly attached to a small, very electronegative, oxygen atom.
- Oxygen is very electron-rich and hydrogen is very electron-deficient.
- The electron-deficient hydrogen atom attracts a lone pair from an oxygen atom on a neighbouring water molecule.
- Hydrogen bonds are stronger than dipole–dipole forces, when comparing molecules of similar size.

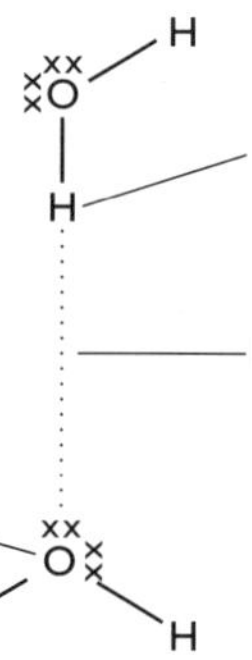

Other examples include ammonia, hydrogen fluoride and methanol.

States of matter

You need to be able to explain the behaviour of gases, liquids and solids in terms of the particles, their motion and the forces acting between them. You also need to be able to explain the energy changes involved in changing states.

The properties of solids

- Particles vibrate about a fixed position.
- There is a high degree of order.
- Solids are incompressible due to the small spaces between particles.
- Solids have a fixed shape and volume.
- The forces holding together a solid vary, but are usually higher than those in the liquid and gas state of the same substance.

An increase in temperature gives the particles more energy and the solid expands. At the melting point, the temperature remains constant because the energy is used to break the forces of attraction. This is called the **latent heat of fusion**.

The properties of liquids

- Liquids are intermediate in order between a solid and a gas, but closer to a solid.
- Liquids are slightly compressible because particles are a little further apart — the separation of particles in liquids is only slightly more than in solids.
- Liquids have no fixed shape but a fixed volume.
- Forces are still present between the particles, but are weaker than in the solid, so particles can move throughout the bulk of the liquid.

If the temperature increases, the kinetic energy of the particles in the liquid increases until the boiling point. The temperature remains constant when the remaining forces of attraction are broken. The energy supplied is called the **enthalpy of vaporisation**.

The properties of gases

- Gas particles move with rapid, random motion.
- Gases have a high degree of disorder or are chaotic.
- Gases have no fixed shape or volume.
- Gases are easy to compress because the particles are far apart.
- The forces of attraction in gases are *negligible*, particularly at low pressures and high temperatures when the particles are very far apart.

Types of crystal

You are expected to be able to draw simple sketches (not of a textbook standard) of the crystal structures. You should also be able to *state* and then *explain* the physical properties of the crystals in terms of structure and bonding. When discussing physical properties, remember three key points — **melting points**, **solubility** and **conductivity**.

Crystals are solids with particles organised in a regular structure. They are classified according to the type of bonding between the particles. There are four types — remember 'I and the three Ms':

- **i**onic
- **m**etallic
- **m**olecular
- **m**acromolecular

Ionic crystals

Example: sodium chloride, NaCl

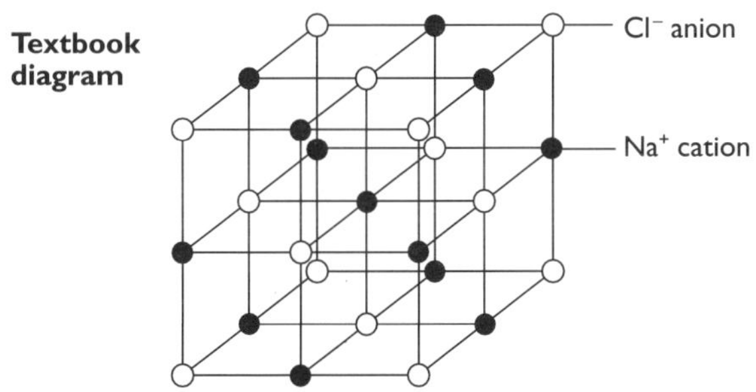

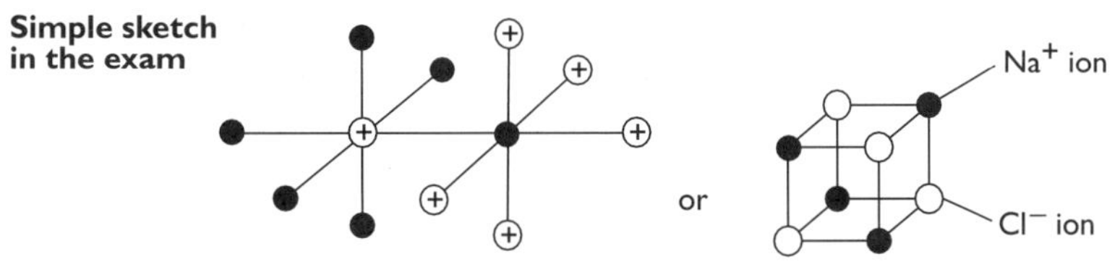

Structure

Sodium chloride is a **giant**, **ionic**, **three-dimensional lattice** of oppositely charged ions with strong electrostatic forces between these ions.

Properties

- High melting point, because of strong electrostatic forces between ions.
- Soluble in polar solvents, because ions interact with water molecules and the lattice breaks down.
- Insulator in the solid state, because the ions can only vibrate about a fixed position.
- Conducts when molten or when in solution, because the ions are mobile.

Metallic crystals

Example: sodium, Na

The typical textbook diagram and the simple sketch in the exam are the same. Make sure you show the cations in a **close-packed arrangement**.

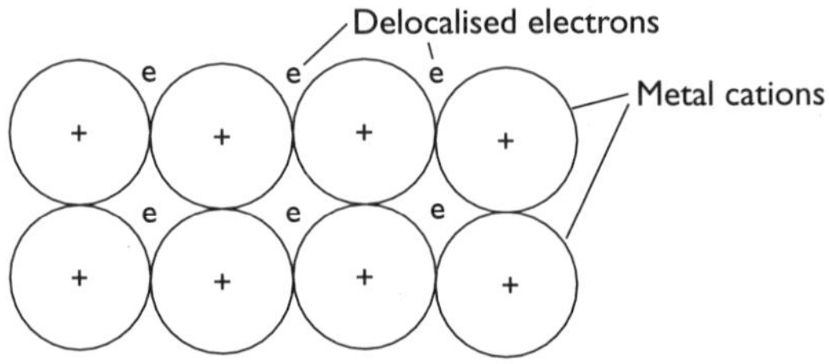

Structure

Metallic crystals consist of a **three-dimensional metallic lattice** of closely packed cations in a sea of delocalised electrons. There are strong forces between the cations and the electrons.

Properties

- Relatively high melting point, because of strong metallic bonds.
- Conducts when solid, due to delocalised electrons.
- Insoluble in water but may react with water.
- Malleable, ductile and reflect light.

Molecular crystals

Example: iodine, I_2

Textbook diagram

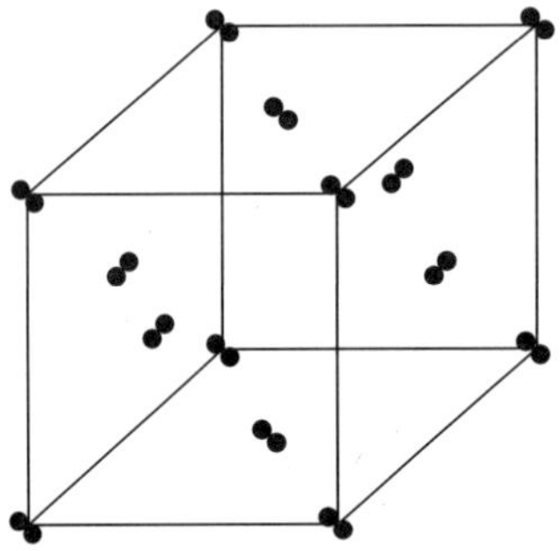

Simple sketch in the exam

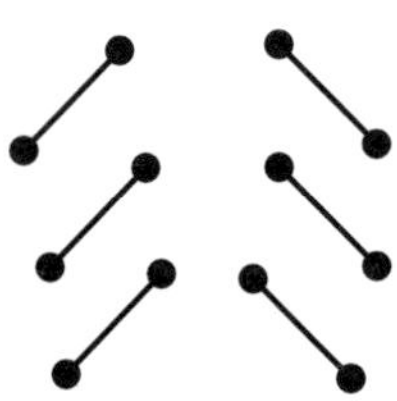

Structure

Iodine has a **molecular lattice** with weak van der Waals forces between the molecules. There are strong covalent bonds between the iodine atoms within each molecule.

Properties

- Iodine sublimes because the weak van der Waals forces between molecules are easily overcome.
- Insoluble in water but soluble in organic solvents.
- Does not conduct because the electrons are localised and not free to move.

Macromolecular (giant atomic) crystals

Examples: diamond, silicon and silicon dioxide

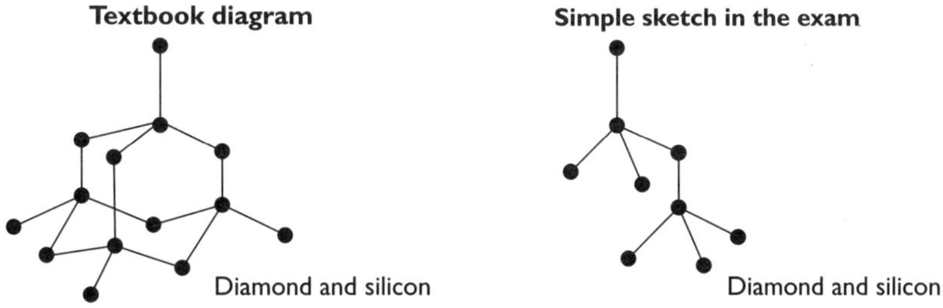

Example: graphite

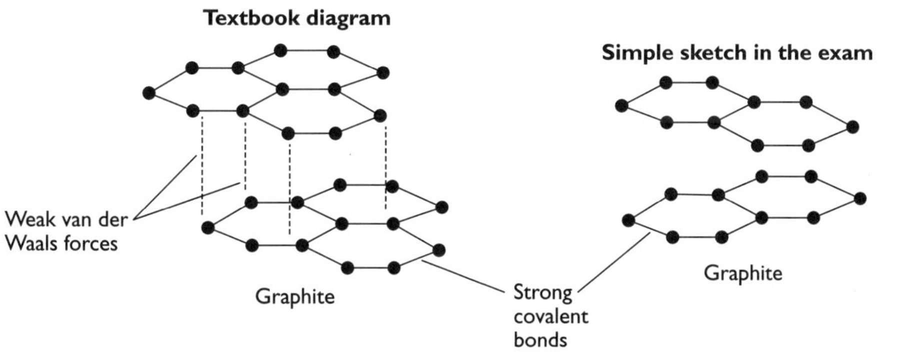

Structure

Diamond	Graphite
Three-dimensional tetrahedral lattice	Hexagons arranged in layers
Carbon has a coordination number of 4	Carbon has a coordination number of 3
All carbon atoms are joined together by strong covalent bonds	Each carbon atom is covalently bonded to three others and weak van der Waals forces hold the layers together

Properties

- Both have very high melting points because the covalent bonds are very strong.
- They are insoluble because the covalent bonds are very strong.
- Diamond does not conduct because the electrons are localised in the bonds.
- Graphite conducts because the lattice contains delocalised electrons which are mobile.

Shapes of simple molecules and ions

Basic principles

- Shape depends upon the number of electron pairs around the central atom.
- Electron pairs will maximise their distance apart in order to minimise repulsion.
- There are bonding pairs (bpr) and lone pairs (lpr).
- The order of repulsion is **lpr/lpr > lpr/bpr > bpr/bpr**.

Basic shapes

There are five basic shapes to remember when only bonding pairs are involved. When drawing any sketch in the exam, make sure you draw the atoms involved and not just the circles.

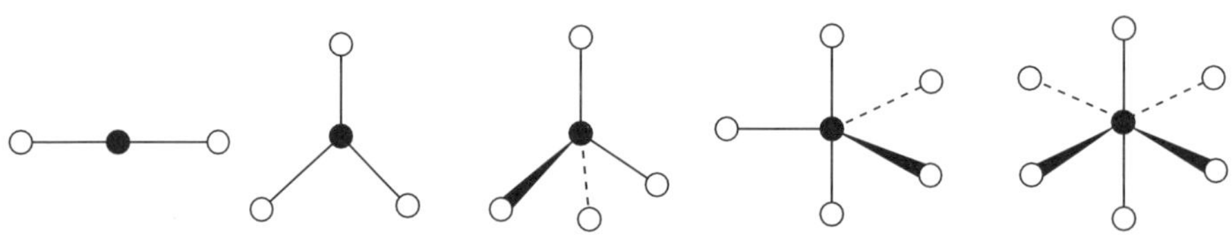

Linear	Trigonal planar	Tetrahedral	Trigonal bipyramidal	Octahedral
2 bpr	3 bpr	4 bpr	5 bpr	6 bpr
180°	120°	109.5°	90 and 120°	90°

The effect of lone pairs

The order of repulsion is lpr/lpr > lpr/bpr > bpr/bpr, so lone pairs reduce the bond angle.

Example: molecules with four electron pairs

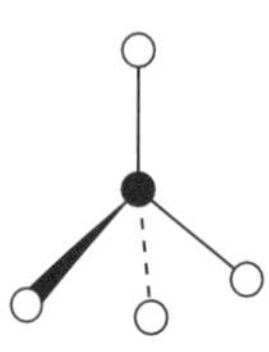

4 bpr
0 lpr
109.5°
Tetrahedral

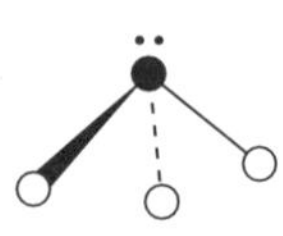

3 bpr
1 lpr
107°
Pyramidal

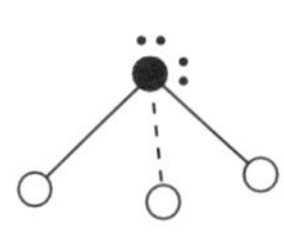

2 bpr
2 lpr
104.5°
Bent planar

An easy way to predict the bond angles is to start with the tetrahedral shape and then take 2.5° off for every lone pair present around the central atom.

A method for determining shapes of molecules

You are required to be able to predict the shapes of molecules and ions that contain only single covalent bonds. The following method works for all the molecules and ions on the AQA specification.

Method	Examples				
	BF_3	SF_6	PCl_3	PCl_4^+	PCl_6^-
Count the total number of electrons in the outer shell of the central atom	3	6	5	5	5
Add 1 if negative ion and subtract 1 if positive ion	0	0	0	–1	+1
Add 1 for each bonded atom	3	6	3	4	6
Total the number of electrons	6	12	8	8	12
Divide by 2 to obtain the number of electron pairs	3	6	4	4	6
Count the number of bonded atoms — the extra pairs are lone pairs	3 bpr	6 bpr	3 bpr 1 lpr	4 bpr	6 bpr
Shape	Trigonal planar	Octahedral	Trigonal pyramidal	Tetrahedral	Octahedral
Bond angle	120°	90°	107°	109.5°	90°

Predicting the shapes of more complex molecules

Do not panic when you are faced with questions of this type; just apply the basic principles set out above. Let us try it using the examples of xenon tetrafluoride (XeF_4) and sulphur tetrafluoride (SF_4).

XeF_4

Xenon (Xe) has 8 outer electrons. Add another 4 for the bonded fluorine atoms and divide by 2 to produce 6 electron pairs; so the molecule is based on an octahedral shape. There are 4 bonding pairs and 2 lone pairs, so the shape is **square planar**. The lone pairs are the maximum distance apart.

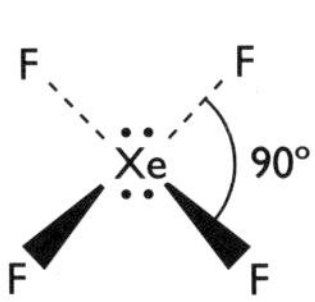

XeF_4

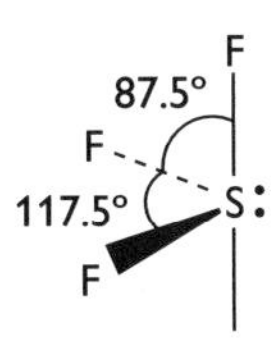

SF_4

SF_4

Sulphur (S) has 6 outer electrons. Add another 4 for the bonded fluorine atoms and divide by 2 to produce 5 electron pairs; so the molecule is based on a **trigonal bipyramidal** shape. However, the lone pair distorts the usual bond angles.

Periodicity

Classification of elements in *s*, *p* and *d* blocks

You need to be able to classify an element as *s*, *p* or *d* block, according to its position in the periodic table.

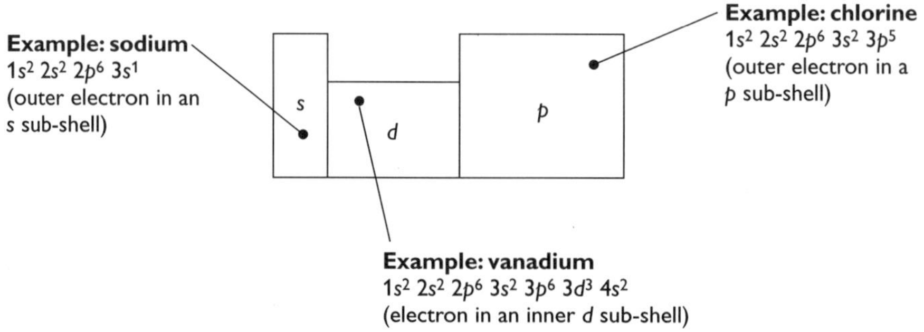

Properties of the elements of period 3

You need to be able to state and then explain, in terms of structure and bonding, the trend in the physical properties of the elements sodium to argon.

The examiner will ask you to state the trend, for 1 mark, and then to explain it, usually for an additional 3 marks. If your statement of the trend is incorrect, you will lose all the marks for the explanation, so it is essential that you can describe every trend in the physical properties of the elements.

Statement

You have to be able either to make a statement or to draw a graph describing the trend in the physical properties, as shown below.

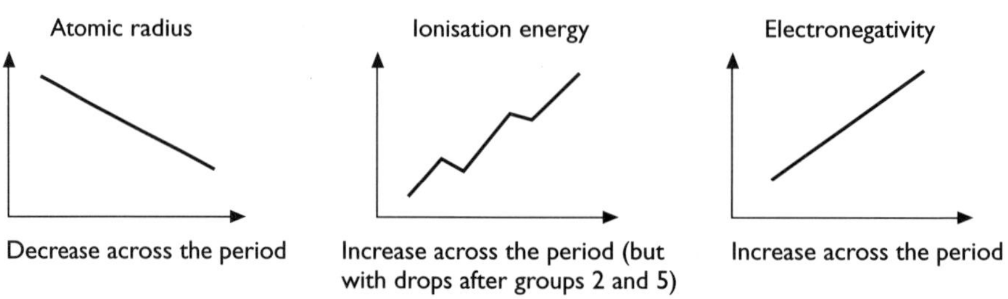

Explanation of trend

You can use the same answer for all three physical properties shown above, except for a slight change in the last sentence.

Atomic radius

- The number of protons (or nuclear charge) increases.
- The shielding stays the same (or electrons enter the same shell).
- The attraction for the electrons is greater, so *the atoms shrink.*

Ionisation energy

- The number of protons (or nuclear charge) increases.
- The shielding stays the same (or electrons enter the same shell).
- The attraction for the electrons is greater, so *the outer electron is harder to lose.*

Electronegativity

- The number of protons (or nuclear charge) increases.
- The shielding stays the same (or electrons enter the same shell).
- The attraction for *the electron density in a covalent bond increases.*

The trend in melting point

Questions on this topic occur frequently. You should to be able to describe in detail

the trend in melting point across period 3 and then relate this trend to the structure and bonding found in the elements.

The trend can be divided into three main sections, as shown on the graph below.

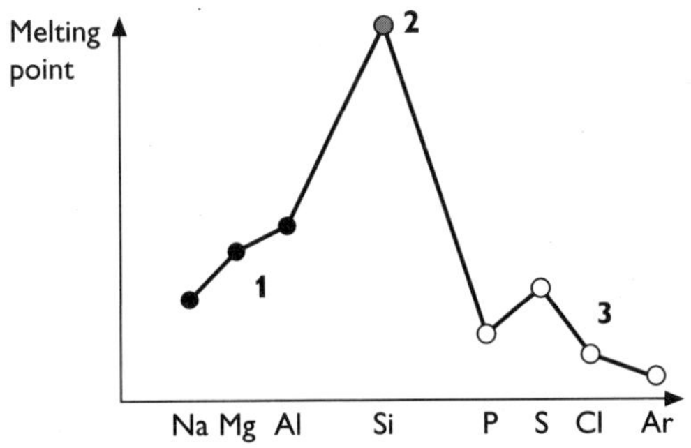

Section 1 of the graph

- The elements sodium (Na), magnesium (Mg) and aluminium (Al) are all metallic.
- They each consist of a giant lattice of cations in a sea of delocalised electrons.
- Melting point increases because metallic bond strength increases.
- Bond strength increases because the nuclear charge of the cations increases; the cations are smaller and there are more electrons per cation, which all contribute to an increased attraction.

Section 2 of the graph

- Silicon (Si) has a very high melting point.
- It is macromolecular.
- Silicon atoms are arranged in a tetrahedral fashion, in a giant lattice similar to diamond.
- Strong covalent bonds link all the silicon atoms together.

Section 3 of the graph

- Phosphorus (P_4), sulphur (S_8) and chlorine (Cl_2) are simple covalent molecules.
- Their melting points are relatively low because only weak van der Waals forces exist between the molecules.
- These van der Waals forces increase with M_r or molecular size.
- Sulphur (S_8) has highest melting point of this section and argon has the lowest.
- Argon (Ar) exists as individual atoms, which are not very polarisable, so the van der Waals forces are very weak.

Group 2

The elements studied are beryllium (Be), magnesium (Mg), calcium (Ca), strontium (Sr) and barium (Ba).

Trends in physical properties

You need to be able to state and then explain the trends in four physical properties of the group 2 elements. It is essential that your statement of the trend is correct because if an incorrect statement is made, you will lose all the marks for the explanation section.

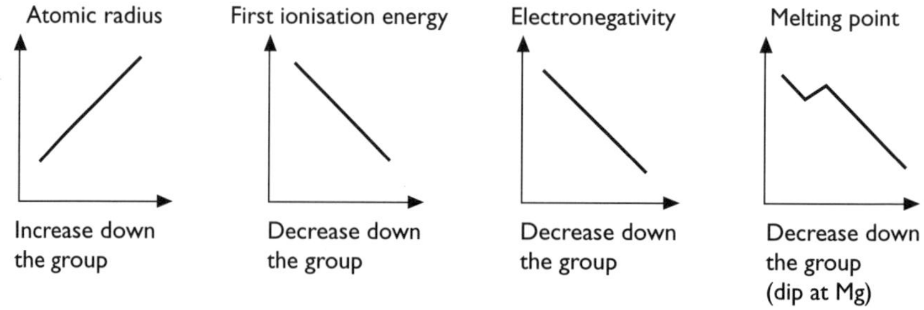

You can use the same answer for the first three properties shown above, except for a change to the last sentence.

Atomic radius

- More shells of electrons are present.
- There is more shielding (this outweighs the increase in protons).
- There is less attraction, so the *atoms increase in size*.

First ionisation energy

- More shells of electrons are present.
- There is more shielding (this outweighs the increase in protons).
- There is less attraction, so the *electron is more easily lost*.

Electronegativity

- More shells of electrons are present.
- There is more shielding (this outweighs the increase in protons).
- There is less attraction *for the electron pair in the covalent bond*.

Melting point

- All elements in group 2 exhibit metallic bonding.

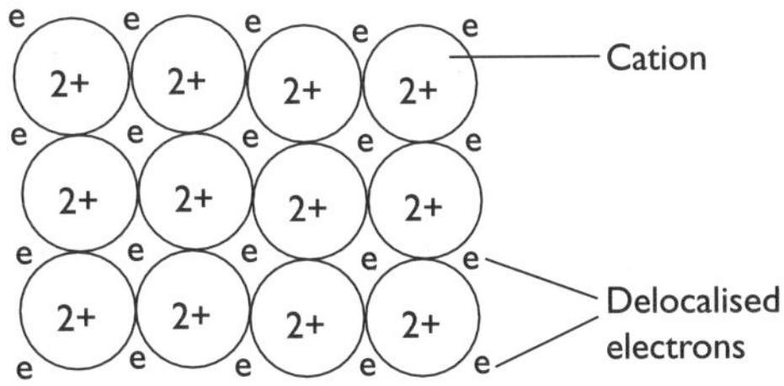

- Cation size increases going down group 2.
- Charge density of the cations decreases down the group.
- There is less attraction for delocalised electrons.
- The metallic bonds are weaker, so the melting points are lower.
- The dip at magnesium is because the metals have different crystal structures and this also affects the melting points.

Reactivity of group 2 metals

Reactivity of the metals increases down the group.

- Beryllium does not react with water.
- Magnesium reacts with steam to produce the oxide and hydrogen:

  $Mg + H_2O \longrightarrow MgO + H_2$
- Calcium, strontium and barium react with cold water to produce the hydroxide and hydrogen:

  $Ba + 2H_2O \longrightarrow Ba(OH)_2 + H_2$

Solubility of group 2 compounds

Trying to remember the trends in the solubilities of the group 2 compounds is difficult, particularly when the topic comes at the end of Module 1. Build up the picture in your mind using the following points:

- Only hydroxides and sulphates have to be remembered.
- The trends for one are the opposite of those for the other.
- Remember that $BaSO_4$ is insoluble — this allows the trends to be deduced.

Metal	OH	SO_4^{2-}
Be	Least soluble, $Be(OH)_2$	Most soluble
Mg	↑	↓
Ca		
Sr		
Ba	Most soluble	Least soluble, $BaSO_4$

Testing for a sulphate

The trend in solubilities can be used to test for a sulphate. If a solution of acidified barium chloride (or nitrate) is added to a sulphate solution, a *white precipitate* is formed.

$$Ba^{2+} + SO_4^{2-} \longrightarrow BaSO_4$$

The acid is used to prevent the formation of barium carbonate ($BaCO_3$), which is also a white precipitate.

The atypical properties of beryllium

Beryllium compounds show some different properties from other group 2 compounds. You must learn three typical properties of group 2 compounds and then learn the different properties of beryllium. The small size of the beryllium ion is the main reason for these different properties.

Typical properties of group 2 compounds

- They form ionic chlorides. $MgCl_2$ is soluble in water and conducts when molten.
- They have basic hydroxides, which are soluble in acids:
  $$Mg(OH)_2 + 4H_2O + 2H^+ \longrightarrow [Mg(H_2O)_6]^{2+}$$
- They form aqueous ions, which have a coordination number of 6. This means that six water molecules can fit around the magnesium ion — $[Mg(H_2O)_6]^{2+}$.

Atypical properties of beryllium

- Beryllium forms a covalent chloride. $BeCl_2$ is soluble in organic solvents and is a poor conductor when molten.
- Beryllium has an amphoteric hydroxide, which is soluble in acids and alkalis.
  $$Be(OH)_2 + 2H_2O + 2H^+ \longrightarrow [Be(H_2O)_4]^{2+}$$
  $$Be(OH)_2 + 2OH^- \longrightarrow Be(OH)_4^{2-}$$
- Beryllium forms aqueous ions, which have a coordination number of 4. This means that four water molecules can fit around the beryllium ion — $[Be(H_2O)_4]^{2+}$.

The reasons for the atypical properties of beryllium

- Beryllium forms a very small cation with a charge of +2.
- The beryllium cation has a very high charge density.
- It is a very polarising cation, which distorts large anions. This introduces covalent character into compounds. It means that the aqueous Be^{2+} ion has a smaller coordination number, because fewer water molecules can fit around a smaller ion.

Questions & Answers

These questions are similar in style and content to those that you can expect in Unit Test 1. Although the number of questions is limited, they have been designed to test the vast majority of key facts and concepts covered in Module 1.

Unit Test 1 is divided into sections A and B. Section A questions are structured, with spaces at the end of each question part for your response. Section B usually contains a longer question, which is also divided into sections. Therefore, it is effectively a long structured question but with no spaces between each section for you to write your answer.

All the questions in this Question and Answer section are structured, without answer spaces. If a question is worth 12 marks, then you should expect to write no more than 12 sentences. The marking scheme identifies 12 key words or phrases that must be present for you to score the 12 marks. The questions are worth 30 marks. In Unit Test 1, it is rare to find one question concentrating on one topic. However, these 30 marks could represent three separate structured questions, each worth 10 marks, from separate papers. The important point is that if you score well in each test, then at least you can be confident that you understand the topic.

Record your scores and convert them to a percentage and a grade. As a rough guide, you need to score 80% to achieve a grade A, 70% for a grade B and 60% for a grade C.

Examiner's comments

Grade-A answers are provided for all questions. These are followed by examiner's comments (preceded by the icon *e*), which point out common errors. They also suggest alternative answers that are acceptable and ways of remembering key points.

Question 1

Atomic structure and mass spectrometry

(a) Describe, in terms of charge and mass, the properties of protons, neutrons and electrons. Explain fully how these particles are arranged in an atom of $^{20}_{10}Ne$. (6 marks)

(b) Two other isotopes of neon exist — $^{21}_{10}Ne$ and $^{22}_{10}Ne$. Explain the meaning of the term *isotopes*. (2 marks)

(c) The three isotopes of neon can be separated in a mass spectrometer. Explain how this is possible, by describing the various parts of the mass spectrometer and by discussing the principles of operation of each part. (14 marks)

(d) The simplified mass spectrum of naturally occurring neon is shown below.

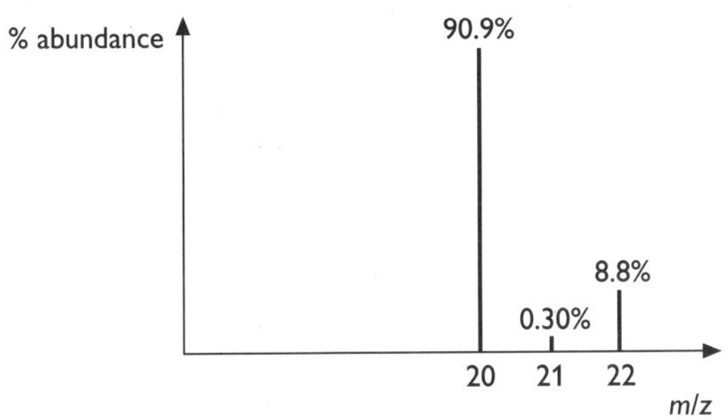

Assuming that all the peaks are produced by ions with a single positive charge, what is the relative atomic mass of neon? (3 marks)

(e) Why is it possible for the mass spectrum of neon to show peaks at 10, 10.5 and 11? (2 marks)

(f) Another element was analysed in the mass spectrometer and the following data were produced.

m/z value	134	135	136	137	138
Relative intensity	1	2.5	3.5	5.5	33

Use the data to calculate the average relative atomic mass and identify the unknown element. (3 marks)

Total: 30 marks

Grade-A answer to Question 1

(a) The properties of charge and mass are summarised in the following table.

Particle	Relative mass	Relative charge	
Proton	1	+1	✓
Neutron	1	0	✓
Electron	1/1840	−1	✓

The nucleus contains ten protons and ten neutrons ✓ and this is surrounded by electrons arranged in shells ✓ with an electron arrangement of $1s^2\,2s^2\,2p^6$ ✓.

e Do not be afraid to summarise information in tables. The mass of an electron can be described as negligible.

(b) Isotopes are elements with the same atomic number ✓ but a different mass number due to differing numbers of neutrons ✓.

e You need to learn definitions so that they are presented concisely. 'Same proton number' instead of 'same atomic number' would also be acceptable.

(c) The gaseous ✓ sample of neon is introduced into the apparatus, which provides a vacuum ✓ for the ions to travel through. An electron gun ✓ is used to bombard the atoms with electrons and this knocks out another electron ✓ from each atom to produce positive ions ✓ (cations).

The cations are accelerated ✓ by an electric field ✓ and passed through a slit ✓so that they are focused into a beam of ions.

The ions are deflected ✓ by passing them through a magnetic field ✓. The amount of deflection depends upon the mass-to-charge (m/z) ratio ✓ of each ion. The ion with the smallest m/z value is deflected the most ✓.

The ions then pass through the apparatus to the detector ✓. When the ions hit the detector, a small current ✓ is produced, which is amplified and sent to a computer where it is recorded and displayed ✓ as a chart. The magnetic or electric field can be adjusted to collect ions of different m/z ratios ✓.

e There are 16 scoring points ticked in this answer, but only 14 marks available for this question. This is quite common on longer questions and you need only score 14 of the 16 points to achieve full marks. Divide your answer into four short paragraphs, include around four key points in each one and do not include too much detail. Your answer should be in short sentences because the mark scheme will be summarised as key statements or key words. One equation that could be included to describe the ionisation process is:

$$Ne(g) + e^- \longrightarrow Ne^+(g) + 2e^-$$

This would score 4 marks, because it implies gaseous atoms, bombardment with electrons, electron loss and production of positive ions.

(d) $A_r = \dfrac{(20 \times 90.9) + (21 \times 0.3) + (22 \times 8.8)}{100}$ ✓✓ = 20.179 = 20.2 ✓

When calculating the average relative atomic mass, make sure that you show your working and that you express your answer to three significant figures.

(e) Sometimes in the ionisation process, two electrons can be removed ✓ to produce cations with a +2 charge. These ions have half the mass-to-charge ratio of the singly charged ions ✓.

The key concept to understand in mass spectrometry is that deflection depends upon the mass-to-charge ratio (*m*/*z*). If two electrons are removed, the charge is +2, so the mass is divided by 2 and these ions require a smaller magnetic field to deflect them.

(f) $$A_r = \frac{(134 \times 1) + (135 \times 2.5) + (136 \times 3.5) + (137 \times 5.5) + (138 \times 33)}{45.5} \checkmark$$

$= 137.4725275$

$= 137.5$ ✓

The unknown element is barium ✓.

A common mistake is to divide by 100 rather than by the total intensity (45.5). Doing this gives an incorrect answer of 62.55 and the identity of the element as copper.

Electron arrangements and ionisation energies

The first ionisation energies of the elements sodium to silicon are given below.

Element	Na	Mg	Al	Si
First ionisation energy/kJ mol^{-1}	494	736	577	786

(a) Explain the meaning of the term *first ionisation energy*. (3 marks)
(b) Explain why magnesium has a higher first ionisation energy than sodium. (3 marks)
(c) State the full electron configurations of magnesium and aluminium. (2 marks)
(d) Explain why the first ionisation energy of aluminium is less than that for magnesium. (2 marks)
(e) Predict a value for the first ionisation energy of phosphorus and explain how you estimated your answer. (2 marks)
(f) State the full electron configuration of phosphorus and sulphur. (2 marks)
(g) Explain why the first ionisation energy of sulphur is less than that of phosphorus. (2 marks)
(h) State and explain the trend in first ionisation energies going down group 2. (4 marks)
(i) Write two separate equations, including state symbols, to describe the first and second ionisation energies of barium. Explain why the second ionisation energy of barium is higher than the first ionisation energy and why the third ionisation energy of barium is significantly higher than the second. (7 marks)
(j) Explain why the second ionisation energy of sodium is higher than the first ionisation energy of neon. (3 marks)

Total: 30 marks

Grade-A answer to Question 2

(a) The enthalpy change when one mole ✓ of gaseous atoms ✓ is converted into 1 mole of gaseous cations, with each atom losing an electron ✓.

e Try to be concise with your definition — remember three key points for 3 marks. It is worth including the general equation $X(g) \longrightarrow X^+(g) + e^-$, because this would be given some credit if your definition was not very clear.

(b) Magnesium has a higher number of protons in the nucleus ✓, but the amount of shielding remains the same ✓, and so the attraction for the outer electrons increases ✓.

e An alternative answer is that magnesium has a higher nuclear charge and the electrons are entering the same shell, so there is greater attraction. The examiner is looking to see if you understand the concepts and although key words are written in mark schemes, other words are acceptable.

(c) Magnesium is $1s^2\,2s^2\,2p^6\,3s^2$ ✓ and aluminium is $1s^2\,2s^2\,2p^6\,3s^2\,3p^1$ ✓.

Sometimes students give the answer $[Ne]3s^2$, but as this is not the *full* electron configuration, it is not acceptable. You must also remember to use *superscript* to show the number of electrons — $1s_2\,2s_2\,2p_6\,3s_2$ is definitely not acceptable.

(d) The electron in aluminium is removed from a 3*p* sub-shell ✓, which is higher in energy ✓ than the 3*s* sub-shell.

Many answers discuss the shielding of the *p* sub-shell by the *s* sub-shell, which is not what the examiner is looking for.

(e) The value for phosphorus will be about 995 kJ mol^{-1} ✓. There is a regular increase from Al to P, so assuming the difference between Al and Si is the same as between Si and P, 209 + 786 = 995 ✓.

Predictive questions are very common and quite often you will be presented with a graph to complete. These are easy questions, provided that you remember the pattern for both period 2 and 3 is 2, 3, 3 followed by a big drop.

(f) Phosphorus is $1s^2\,2s^2\,2p^6\,3s^2\,3p^3$ ✓ and sulphur is $1s^2\,2s^2\,2p^6\,3s^2\,3p^4$ ✓.

Remember to give the *full* electron configurations of the phosphorus and sulphur atoms. A useful way of showing whether one of the *p* orbitals contains a pair of electrons is to write out the electron configurations to show the 3*p* orbitals as X, Y and Z. So phosphorus is $1s^2\,2s^2\,2p^6\,3s^2\,3p_X^1\,3p_Y^1\,3p_Z^1$ and sulphur is $1s^2\,2s^2\,2p^6\,3s^2\,3p_X^2\,3p_Y^1\,3p_Z^1$.

(g) Sulphur contains a pair of electrons ✓ in one of the orbitals of the 3p level — [↑↓] [↑] [↑] — and this pair suffers repulsion ✓ — making the electron easier to remove compared with phosphorus, which has three unpaired electrons in the three orbitals — [↑] [↑] [↑].

A common phrase used by candidates is 'spin-pair repulsion'. However, to convince the examiner you understand the concept, you need a little more detail.

(h) The first ionisation energy decreases ✓ going down group 2, because there is an increase in the number of shells ✓ and so there is more shielding ✓ and hence less attraction for the outer electrons ✓.

Many questions involving a trend start with the words 'state and explain'. If your statement is incorrect it will be marked as a chemical error and you will not receive any marks for the explanation. In this case, therefore, a wrong statement would cost 4 marks.

(i) The first ionisation energy:

$Ba(g) \longrightarrow Ba^+(g) + e^-$ ✓✓

The second ionisation energy:

$Ba^{+}(g) \longrightarrow Ba^{2+}(g) + e^{-}$ ✓

The second ionisation energy is higher than the first because there is still the same number of protons ✓ and the ion is now positive and attracts the remaining electrons more strongly ✓.

The third ionisation energy is significantly higher than the second, because the electron is now removed from a new inner shell, closer to the nucleus ✓, which experiences less shielding ✓ and so there is more attraction.

e You need to include all the correct state symbols to gain the third mark in this question. A common mistake in explaining the increase in successive ionisation energies is to state that the attraction is greater because the nuclear charge or the number of protons is increasing, when in fact they are constant.

(j) Consider the electron configuration of neon and a sodium ion: Ne $1s^2\,2s^2\,2p^6$ and Na^{+} is $1s^2\,2s^2\,2p^6$.

Both involve removing an electron from the $2p$ level ✓, but sodium has 11 protons in its nucleus, whereas neon has only 10 ✓. The higher nuclear charge means a greater attraction to the electron in the $2p$ level ✓.

e The last question is difficult and would be used to discriminate at the top end of the ability range. When discussing questions of this type, always look to see how many protons are present in the nucleus, how many electrons are surrounding the nucleus and the energy level of the shell where the electron is removed.

Question 3

Amount of substance (I)

The gas constant R = 8.31 $JK^{-1} mol^{-1}$
The Avogadro constant L = 6.023×10^{23}

(a) (i) Define the term *relative atomic mass*. (2 marks)
(ii) Using the Avogadro constant, calculate the mass of an atom of carbon-12. (1 mark)

(b) An organic liquid is analysed in a mass spectrometer and found to have a relative molecular mass of 88. It contains 54.5% carbon, 36.4% oxygen and 9.1% hydrogen. Calculate the empirical formula and deduce the molecular formula of the compound. Write an equation for the complete combustion of this compound to produce carbon dioxide and water. (6 marks)

(c) An impure sample of aminosulphonic acid, H_2NSO_3H, was analysed by a student using the following method. 2.50 g of the impure sample was dissolved in 250 cm^3 of water. A 25.00 cm^3 sample was pipetted into a conical flask and then titrated against a 0.100 mol dm^{-3} solution of sodium hydroxide. 24.90 cm^3 was required.

$$H_2NSO_3H + NaOH \longrightarrow H_2NSO_3Na + H_2O$$

Calculate the percentage by mass of aminosulphonic acid in the impure sample. (7 marks)

(d) (i) State the ideal gas equation and show how it can be used to calculate the relative molecular mass of a gas.
(ii) Use this expression to calculate the relative molecular mass of propane. 1.665 g of propane occupies a volume of 1 dm^3 at 323 K and 101 kPa. (6 marks)

(e) Calculate the volume of carbon dioxide that would be evolved when 4 g of sodium carbonate reacts completely with excess dilute hydrochloric acid at 30 °C and 100 kPa.

$$Na_2CO_3 + 2HCl \longrightarrow 2NaCl + CO_2 + H_2O$$ (5 marks)

(f) In the industrial extraction of iron, iron(III) oxide is reduced using carbon monoxide. How many tonnes of iron can be obtained from the complete reduction of 80 tonnes of iron(III) oxide with excess carbon monoxide?

$$Fe_2O_3 + 3CO \longrightarrow 2Fe + 3CO_2$$ (3 marks)

Total: 30 marks

■ ■ ■

Grade-A answer to Question 3

(a) (i) The relative atomic mass is the average mass of an atom ✓ relative to $\frac{1}{12}$ of carbon-12✓, where carbon is 12.000.

(ii) One mole of carbon has a mass of 12.000 g

$$\frac{12}{6.023 \times 10^{23}} = 1.992 \times 10^{-23} \text{ g ✓}$$

You need to be able to recall this definition very precisely. You could express it as a formula:

$$A_r = \frac{\text{average mass of an atom}}{\frac{1}{12}\text{ the mass of carbon-12}}$$

The calculation depends upon your knowledge that 1 mole of any substance contains 6.023×10^{23} particles.

(b)

	Carbon	Hydrogen	Oxygen	
% composition	54.5	9.1	36.4	
A_r	12	1	16	✓
Divide by A_r	4.54	9.1	2.275	✓
Divide by smallest	2	4	1	

The 2:4:1 ratio shows that the empirical formula is C_2H_4O ✓. The compound has a relative molecular mass of 88, so its molecular formula is $C_4H_8O_2$ ✓.

$C_4H_8O_2 + 5O_2$ ✓ $\longrightarrow$ $4CO_2 + 4H_2O$ ✓

It is essential that you show your working. If you make a slip, then consequential marks can be awarded for the correct working out. The specification requires you to be able to write equations for unfamiliar reactions. A common mistake is to try to balance the equation without including oxygen on the left-hand side.

(c) Moles of NaOH $\frac{MV}{1000} = \frac{0.1 \times 24.9}{1000} = 2.49 \times 10^{-3}$ ✓

Moles of H_2NSO_3H in 25 cm^3 = 2.49×10^{-3} ✓

Moles of H_2NSO_3H in 250 cm^3= 2.49×10^{-2} ✓

Formula mass of H_2NSO_3H = 97 ✓

Mass of $H_2NSO_3H = 97 \times 2.49 \times 10^{-2} = 2.4153$ g ✓

% purity = $\frac{2.4153 \times 100}{2.5}$ ✓ = 96.6% ✓

Many candidates would not be able to complete this calculation. However, credit is given for completing separate sections. For example, M_r = 97 on its own would score 1 mark. If you failed to scale up by a factor of 10 in stage 3, your answer would be 9.66%, scoring 5 marks. To score all 7 marks, the final answer must be given to three significant figures.

(d) (i) $pV = nRT$ ✓ and $n = \frac{m}{M_r}$ ✓

so

$$pV = \frac{mRT}{M_r}$$

rearranging

$$M_r = \frac{mRT}{pV}$$ ✓

(ii) $M_r = \dfrac{1.665 \times 8.31 \times 323}{101 \times 10^3 \checkmark \times 1 \times 10^{-3} \checkmark} = 44.25 \checkmark$

In part (i), it is essential to show the three stages involved in deriving the final expression. If the final expression is shown, it assumes that you have understood the basic concept, so the second mark is also scored. If the correct answer of 44.25 is shown for part (ii), then 3 marks are awarded without the need for any working out. However, you are advised to show your working out just in case you make a careless slip. You should always convert kPa to Pa and dm^3 to m^3, but in this calculation you would obtain the correct answer of 44.25 even if you failed to convert both of them — consider yourself very lucky!

(e) Moles of $Na_2CO_3 = \dfrac{m}{M_r} = \dfrac{4}{106} \checkmark = 0.037 \checkmark$

Moles of $CO_2 = 0.037 \checkmark$

$V = \dfrac{nRT}{p} = \dfrac{0.037 \times 8.31 \times 303}{100 \times 10^3} \checkmark = 9.5 \times 10^{-4}\,m^3 \checkmark$ (or $0.95\,dm^3$ or $950\,cm^3$)

There are four definite stages in this calculation. The first mark is for the M_r and the second is for the number of moles of Na_2CO_3. You need to realise that the ratio of $Na_2CO_3 : CO_2$ is 1 : 1, as shown in the equation, to achieve the third mark. The fourth mark is for using the rearranged ideal gas equation. Common mistakes include not converting °C to K (by adding 273) and not converting kPa to Pa. The final answer will only gain a mark if the *units* are given.

(f) $Fe_2O_3 \longrightarrow 2Fe$

160 g ✓ ⟶ 2(56) = 112 g ✓

80 g ⟶ 56 g

80 tonnes ⟶ 56 tonnes ✓

This reacting masses question is relatively easy compared with some of the other calculations in this section. Marks are awarded for calculating the M_r of the compounds, realising that it is a 1:2 ratio in the equation and scaling down to produce the correct answer, which must include the correct units.

Amount of substance (II)

The gas constant R = 8.31 $JK^{-1} mol^{-1}$

(a) (i) Define the term *relative molecular mass.* (2 marks)

(ii) The number of molecules in one mole of carbon dioxide is 6.023×10^{23}. What is the name given to this number of molecules? (1 mark)

(iii) Calculate the number of molecules in 2.20 g of carbon dioxide. (2 marks)

(b) (i) Define the terms *empirical formula* and *molecular formula.* (4 marks)

(ii) An organic compound contains 58.8% carbon, 9.8% hydrogen and 31.4% oxygen. It has a relative molecular mass of 204. Calculate the empirical formula and deduce the molecular formula of the compound. (4 marks)

(c) Magnesium reacts with dilute hydrochloric acid to produce magnesium chloride and hydrogen:

$$Mg(s) + 2HCl(aq) \longrightarrow MgCl_2(aq) + H_2(g)$$

(i) Calculate the volume of 0.5 mol dm^{-3} hydrochloric acid needed to completely react with 1.5 g of magnesium ribbon. (3 marks)

(ii) Calculate the volume of hydrogen gas evolved at 30°C and 101 kPa when 1.5 g of magnesium ribbon completely reacts with excess dilute hydrochloric acid. (3 marks)

(d) Magnesium carbonate decomposes to produce magnesium oxide and carbon dioxide:

$$MgCO_3(s) \longrightarrow MgO(s) + CO_2(g)$$

Calculate the mass of magnesium oxide formed when 10.0 g of magnesium carbonate is completely decomposed. (3 marks)

(e) Calculate the relative molecular mass of an unknown gas, which has a density of 1.798 g dm^{-3} (i.e. 1.798 g has a volume of 1 dm^3) at 298 K and 10 kPa. (4 marks)

(f) You are provided with two separate solutions of 0.100 mol dm^{-3} sulphuric acid.

(i) Calculate the mass of zinc needed to completely react with 50.0 cm^3 of a 0.100 M solution of sulphuric acid.

$$Zn + H_2SO_4 \longrightarrow ZnSO_4 + H_2O$$ (2 marks)

(ii) Calculate the volume of 0.200 mol dm^{-3} sodium hydroxide solution needed to completely react with 50.0 cm^3 of 0.100 mol dm^{-3} solution of sulphuric acid.

$$2NaOH + H_2SO_4 \longrightarrow Na_2SO_4 + H_2O$$ (2 marks)

Total: 30 marks

■ ■ ■

Grade-A answer to Question 4

(a) (i) $M_r = \dfrac{\text{average mass of a molecule}}{\frac{1}{12} \text{ of the mass of one atom of carbon–12}}$ ✓✓

(ii) 6.023×10^{23} is the Avogadro constant. ✓

(iii) The number of moles of $CO_2 = \frac{2.2}{44} = 0.05$ mol ✓

The number of molecules of $CO_2 = 0.05 \times 6.023 \times 10^{23} = 3.0115 \times 10^{22}$ ✓

e You can give the definition as a sentence or express it as a formula. One mole of any substance contains 6.023×10^{23} particles. This number is the Avogadro constant. The number of particles can be calculated if the mass of the substance is given. Always convert the mass to moles, then multiply by the Avogadro constant. Many candidates find it difficult to enter these numbers into their calculators.

(b) (i) The empirical formula is the simplest whole number ratio ✓ of atoms of each element in a compound ✓. The molecular formula is the actual number of atoms ✓ of each element in a compound ✓.

(ii)

	Carbon	Hydrogen	Oxygen
% composition	58.8	9.8	31.4
A_r	12	1	16
Divide by A_r	4.9	9.8	1.9625 ✓
Divide by smallest	2.5	5	1 ✓
In whole numbers	5	10	2

The 5 : 10 : 2 ratio shows that the empirical formula is $C_5H_{10}O_2$ ✓. The compound has a relative molecular mass of 204, so its molecular formula is $C_{10}H_{20}O_4$ ✓.

e In many questions in the 'Amount of substance' topic, marks can be scored for recalling simple definitions. It is essential that you *learn all your definitions*. The first stage of the calculation involves dividing the percentage mass by the relative atomic mass of the element. A common mistake is to divide the relative atomic mass by the percentage mass. When calculating the empirical formula it is essential that you derive the simplest *whole number* ratio. The molecular formula will always be a whole number multiple of the empirical formula.

(c) (i) Moles of magnesium $= \frac{1.5}{24} = 0.0625$ ✓

Moles of HCl = 2 × 0.0625 = 0.125 ✓

$$n = \frac{MV}{1000}$$

Therefore

$0.125 = \frac{0.5 \times V}{1000}$ and $V = 250\ \text{cm}^3$ ✓

(ii) Moles of magnesium $= \frac{1.5}{24} = 0.0625$ ✓ (only awarded if not gained in part (i))

Moles of hydrogen = 0.0625 ✓

$pV = nRT$

Therefore $V = \frac{nRT}{P} = \frac{0.0625 \times 8.31 \times 303}{101 \times 10^3}$ ✓ $= 1.56 \times 10^{-3}\,m^3$ ✓ (or $1.56\,dm^3$)

If you produce the correct final answers, with the appropriate units, then you score 6 marks for this section. However, it is essential to *show your working*, because careless mistakes can be made. If you make a slip in calculating the number of moles of magnesium, you can still score the next 2 marks in each calculation, provided that your working is correct. Many candidates find part (ii) difficult. Common mistakes include failing to convert the temperature from °C to K and failing to convert kPa to Pa. If you cannot attempt these gas calculations, it is always worth including $pV = nRT$ as a statement because it will probably score 1 of the marks.

(d) $MgCO_3 \longrightarrow MgO + CO_2$
84 g ✓ ⟶ 40 g ✓
10 g ⟶ $40 \times 10/84 = 4.76$ g ✓

The correct answer scores 3 marks. If you have made a mistake, you can still score marks for deducing the M_r of each substance.

(e) $pV = nRT$ ✓ $= \frac{mRT}{M_r}$

Rearranging gives

$$M_r = \frac{mRT}{pV} \checkmark$$

$$M_r = \frac{1.798 \times 8.31 \times 298}{101 \times 10^3 \times 1 \times 10^{-3}} \checkmark = 44.1 \checkmark$$

When attempting this type of question, always start with $pV = nRT$ and rearrange the expression to give $M_r = mRT/pV$. This would probably score 2 marks.

(f) (i) Moles of sulphuric acid $= \frac{MV}{1000} = \frac{0.1 \times 50}{1000} = 5 \times 10^{-3}$ ✓

Moles of zinc $= 5 \times 10^{-3}$

Mass of zinc $= 5 \times 10^{-3} \times 65.4 = 0.327$ g ✓

(ii) Moles of sulphuric acid $= 5 \times 10^{-3}$
1 : 2 ratio in the equation, so moles of NaOH $= 1 \times 10^{-2}$ ✓
Volume of NaOH required, $V, = \frac{n \times 1000}{V} = \frac{1 \times 10^{-2} \times 1000}{0.2} = 50\,cm^3$ ✓

Remember the *three key stages* in this type of calculation:
- Calculate moles of acid using *MV*/1000.
- Use the ratio in the equation, i.e. 1 : 1 or 1 : 2.
- Use $n = m/M_r$ if a solid is involved and $n = MV/1000$ if a solution is involved.

Question 5

Types of bonding

You should revise bonding types and shapes of molecules before attempting this question.

(a) Give an example and then explain the formation and principle features of (i) ionic, (ii) covalent and (iii) metallic bonds using *only* the elements *potassium* and *fluorine* or *any combination* of these two elements. (12 marks)

(b) Name and sketch the shape of water (H_2O) and ammonia (NH_3) molecules, indicating the bond angles and the presence of any lone pairs. (8 marks)

(c) The water and ammonia molecules can both accept an H^+ ion by forming a coordinate bond. Define the term coordinate bond, name and sketch the shape, including the bond angles, of the new species formed in each case and comment on the relative strength of all the N–H bonds in the final species. (10 marks)

Total: 30 marks

Grade-A answer to Question 5

(a) (i) An ionic bond is formed between potassium and fluorine to give KF ✓. Potassium ($1s^2\,2s^2\,2p^6\,3s^2\,3p^6\,4s^1$) loses an electron to become K^+ ✓. The electron is transferred ✓ to the fluorine atom. Fluorine ($1s^2\,2s^2\,2p^5$) gains an electron to become a fluoride ion, F^- ✓. The electrostatic forces of attraction ✓ between the oppositely charged ions lead to the formation of a giant, three-dimensional, ionic lattice ✓.

(ii) A covalent bond is formed between two fluorine atoms to give F_2 ✓. Both fluorine atoms ($1s^2\,2s^2\,2p^5$) need to gain an extra electron ✓ and this is achieved by sharing ✓ an electron pair ✓. This results in the formation of a fluorine molecule, F_2 ✓.

(iii) A metallic bond is formed between atoms of potassium, K ✓. Potassium atoms lose control of the outer electrons to form K^+ cations ✓. The electrons that are lost form a sea of delocalised ✓ electrons ✓. The metal is a three-dimensional lattice ✓ of cations held together in a closely packed ✓ arrangement by their mutual attraction for the delocalised electrons.

e This answer offers 17 scoring points. However, only 12 marks can be scored. You should structure your answer using the question and the mark allocation given. There are three types of bonding to discuss and the question asks for an example in each case. The mark scheme will usually give a maximum of 4 marks for each type of bonding discussed (1 mark for the example and 3 for the explanation). If you score all 7 marks in the first section, you will still only be awarded 4 marks.

(b) The shapes of water and ammonia molecules are shown below.

Bent planar ✓ Trigonal pyramidal ✓

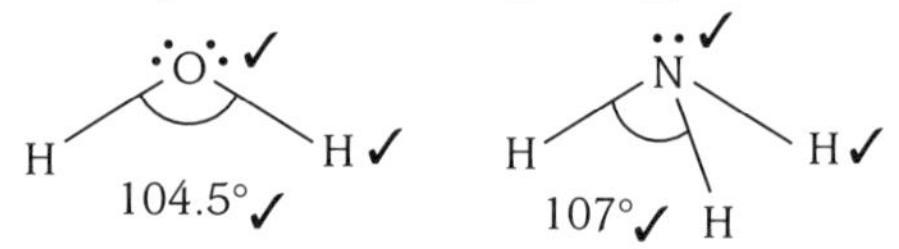

The question and the mark allocation give you an idea of how to structure your answer. When reading through questions in the exam, you should highlight the key words in the question. In this question you would highlight 'name', 'sketch', 'bond angles' and 'lone pairs'. These four key points equate to 4 marks and as you have to discuss two molecules, it is clear how the 8 marks are awarded.

(c) A coordinate bond is defined as a shared pair of electrons ✓ where both electrons have originated from one atom ✓.

Trigonal pyramidal ✓ Tetrahedral ✓

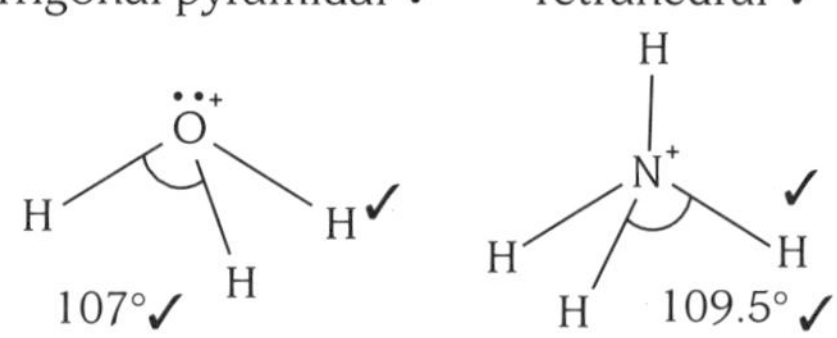

The four N–H bonds in the final species (NH_4^+, ammonium ion) are all exactly the same strength ✓ because they involve the sharing of an electron pair ✓, i.e. a coordinate bond is the same as any other covalent bond.

Repeat the same procedure in this section. Two ions each require a name, a sketch and the bond angles, producing 6 marks. There are two other questions left in this section. 2 marks for the definition and 2 marks for the final explanation give a final total of 10 marks.

It is very common to see a question that relies on your understanding of more than one topic. In this question on bonding, a detailed knowledge of the basic principles of shapes of molecules is also required. This question demonstrates how to use the requirements and the mark allocation to structure your answer.

Bond polarity and polarisation of ions

(a) Define the term *electronegativity*. (2 marks)

(b) State and explain the variation in electronegativity (i) across period 3 and (ii) down group 2. (8 marks)

(c) Using suitable simple molecules as examples, explain the terms *non-polar covalent bond* and *polar covalent bond*. (7 marks)

(d) The structures of chloromethane and tetrachloromethane are shown below.

Both molecules have four bonds arranged in a tetrahedral fashion around the carbon atom. In chloromethane, show the bond polarity of the carbon–chlorine bond by using the symbols δ+ and δ– to indicate the electron-rich and electron-deficient atoms. Suggest why the molecules of tetrachloromethane show no overall polarity. (3 marks)

(e) Caesium fluoride (CsF) is an ionic compound, but lithium iodide is ionic with a large degree of covalent character. Explain the meaning of the term 'covalent character' and why there is a difference between these two compounds. (6 marks)

(f) Indicate which compound in each pair below exhibits the most covalent character. Give a brief reason for your choice.

(i) magnesium chloride and aluminium chloride (2 marks)

(ii) lithium iodide and lithium chloride (2 marks)

Total: 30 marks

Grade-A answer to Question 6

(a) Electronegativity is the ability of an atom to attract the electron density ✓ in a covalent bond ✓.

'The ability to attract an electron pair in a covalent bond' is also an acceptable answer. 'The ability to attract an electron' would score no marks. It is essential that you refer to the covalent bond.

(b) (i) Electronegativity increases ✓ across period 3 because the number of protons in the nucleus is increasing ✓ and shielding remains constant ✓, so there is an increased attraction ✓ for the electron pair in the covalent bond.

(ii) Electronegativity decreases ✓ down group 2 because there are more shells ✓ and more shielding ✓, which outweigh the increased nuclear charge, so there is less attraction ✓ for the electron pair in the covalent bond.

e If the statement is wrong, then this will be marked as a chemical error and no marks will be awarded for the explanation. In part (i), 'an increase in nuclear charge' is an alternative answer to 'an increase in the number of protons', while 'the electrons enter the same shell' is an alternative answer to 'the shielding remains constant'. When discussing electronegativity, always finish off with 'the attraction for an electron pair in a covalent bond'.

(c) A covalent bond is a shared ✓ electron pair ✓. This electron pair is shared equally when both atoms have the same electronegativity ✓, so the bond is non-polar. An example is chlorine ✓.

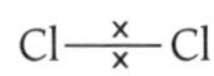

A polar covalent bond occurs when the electron pair is not shared equally ✓. In hydrogen chloride, for example, the chlorine is more electronegative than the hydrogen ✓, so the bond is polar. The chlorine is electron-rich and the hydrogen is electron-deficient ✓.

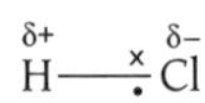

e You need to score 7 marks out of the 8 scoring points shown here. Use simple examples to illustrate your answer. Any diatomic molecule can be used to show a non-polar bond, for example H_2, O_2, F_2, Br_2, I_2. Simple molecules that contain elements of different electronegativities can be used to show a polar covalent bond. Examples include HBr, HI, H_2O and NH_3. The mark for the electron-deficient and electron-rich sites can be scored by labelling the diagram with δ+ and δ–.

(d) The polar bond in chloromethane is shown below.

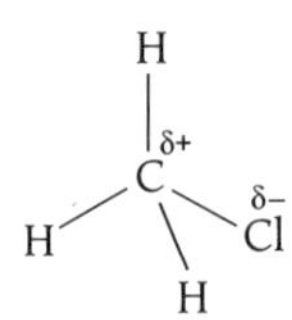

CCl_4 is a symmetrical ✓ molecule, so all the bond polarities cancel out ✓ to give zero polarity.

e Molecules that have polar bonds arranged in a symmetrical fashion have no overall polarity. Another common example is CO_2, a linear molecule with no overall polarity, and SO_2, a bent planar molecule with an overall dipole.

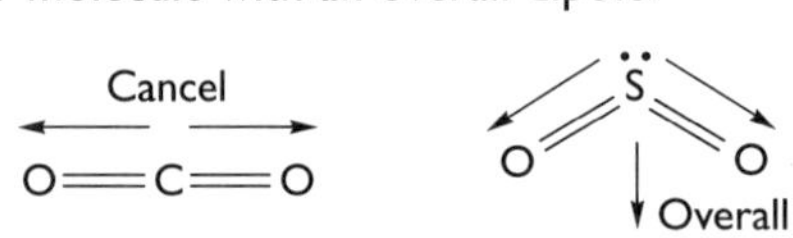

(e) Covalent character occurs in ionic compounds when a cation, with a high charge density ✓, is next to a large polarisable anion ✓. The cation is very polarising ✓ and withdraws electron density from the anion, so that the shape of the anion is distorted ✓. LiI is more covalent in character than CsF because the Li^+ ion is smaller ✓ than the Cs^+ ion, so the cation is more polarising, while the I^- ion is larger ✓ than the F^-, so the anion is more polarisable.

e Using diagrams would help to explain this difficult concept.

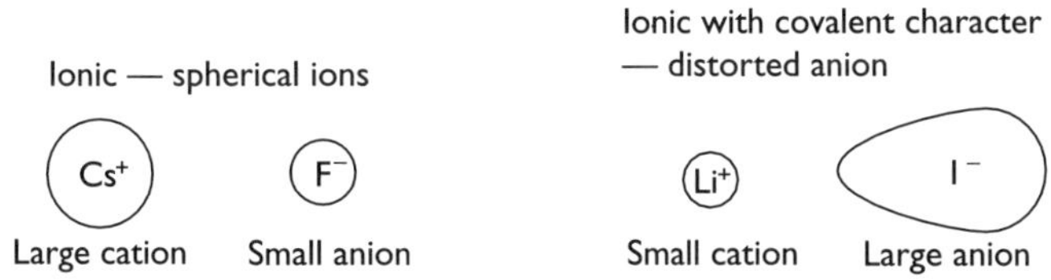

(f) Aluminium chloride ✓ is the more covalent in character because the Al^{3+} ion has a higher charge and a smaller size (a higher charge density) ✓ compared with the Mg^{2+} ion, and so it is more polarising.

Lithium iodide ✓ is more covalent because the I^- ion is larger ✓ and more polarisable than the Cl^- ion.

e Apply the basic principles outlined above. Metal cations increase in charge and decrease in size across any given period, so the order of polarising power is:

$Al^{3+} > Mg^{2+} > Na^+$

Anions increase in size down any group, so the ease of distortion of group 7 anions is:

$I^- > Br^- > Cl^- > F^-$

Intermolecular forces and states of matter

(a) There are three types of intermolecular force: van der Waals forces, dipole–dipole forces and hydrogen bonding. Using an example in each case, explain how each type of intermolecular force arises. (12 marks)

(b) (i) Explain why iodine exists as a solid and chlorine exists as a gas at room temperature and pressure. (2 marks)

(ii) Describe the motion of the particles in solid iodine and iodine vapour. (3 marks)

(iii) Explain why solid iodine can be vaporised by gentle heating. (1 mark)

(c) Sketch a molecule of ammonia and include any lone pairs of electrons. Show how a hydrogen bond forms between this ammonia molecule and a neighbouring ammonia molecule. Indicate the polarity of any atoms involved in the hydrogen bond with the symbols δ+ and δ–. (4 marks)

(d) Which of the molecules shown below does not exhibit hydrogen bonding? Give a brief reason for your answer. (2 marks)

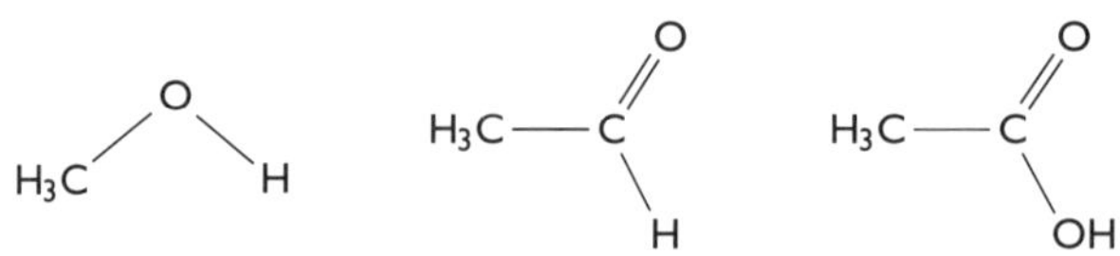

Methanol Methanal Methanoic acid

(e) Indicate which compound in each of the following pairs of substances has the higher boiling point. Give a brief reason for your answer.

(i) CH_4 and SiH_4 (2 marks)

(ii) C_3H_8 and CH_3Cl (2 marks)

(iii) H_2O and H_2S (2 marks)

Total: 30 marks

■ ■ ■

Grade-A answer to Question 7

(a) van der Waals forces attract iodine ✓ molecules to each other. Iodine is a non-polar ✓ molecule, but the electron distribution can become asymmetrical (unequal) ✓ leading to a temporary dipole ✓. This induces an opposite dipole in the adjacent molecule ✓ which is therefore attracted to the first molecule.

Dipole–dipole forces attract hydrogen chloride ✓ molecules to each other. It is a polar ✓ molecule because permanent charge separation ✓ exists between the hydrogen and chlorine atoms. Chlorine is more electronegative than hydrogen ✓, so chlorine is electron-rich and hydrogen is electron-deficient. One dipole attracts the opposite dipole ✓ on a neighbouring molecule.

Hydrogen bonding is found between molecules such as water ✓. It is a special case of a dipole–dipole force ✓. The hydrogen atom is attached directly to an oxygen ✓ atom, meaning the hydrogen becomes so electron-deficient ✓ that it attracts a lone pair from a neighbouring molecule ✓.

There are five scoring points in each of these answers. Four are needed for full marks. You score 1 mark for the example and then any three of the remaining four points give you 3 marks for explanation. You could include *labelled* diagrams to improve your chances of scoring all the points. The examiner will be looking to see if you understand the concept and so will look for certain *key points* in a description or implied in a diagram.

(b) (i) Iodine is a larger ✓ molecule than chlorine, so it has stronger ✓ van der Waals forces.

(ii) In solid iodine, the particles vibrate ✓ about a fixed position ✓, while in iodine vapour, they move with rapid, random ✓ motion.

(iii) Iodine can be vaporised by gentle heating because the van der Waals forces are weak ✓.

There are very often questions on states of matter and most require only simple answers. Most candidates would find this question relatively straightforward.

(c) The hydrogen bonding in ammonia is shown below.

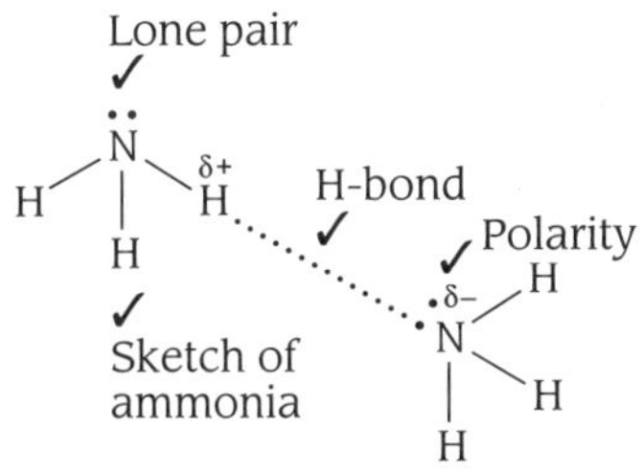

The examiner is looking for a clear sketch of ammonia that shows three N–H bonds and one lone pair on the nitrogen. The shape of the molecule is pyramidal with bond angles of 107°. However, the geometry of ammonia is not being tested, so there are no marks available for this on your sketch. You must then show the hydrogen bond (usually indicated by a dotted line) between the hydrogen of the first molecule and the lone pair on the nitrogen. The final mark is for showing the electron-deficient hydrogen and the electron-rich nitrogen.

(d) Methanal ✓ does not exhibit hydrogen bonding because the hydrogen atom must be attached ✓ directly to the electronegative oxygen atom.

This is an application question that tests understanding of the concept of hydrogen bonding. The key point here is that hydrogen must be attached directly to a small electronegative atom such as oxygen to cause hydrogen bonding between molecules.

(e) (i) SiH_4 ✓ has the higher boiling point because it is a larger molecule with stronger van der Waals forces ✓.

(ii) CH_3Cl ✓ has a higher boiling point because it is a polar molecule, so the dipole–dipole forces ✓ are stronger than the van der Waals forces found between C_3H_8 molecules.

(iii) H_2O ✓ has a higher boiling point because the hydrogen bonding ✓ is stronger than the dipole–dipole forces found in hydrogen sulphide.

e If you understand the basic principles of intermolecular forces, you should be able to apply them to any molecule. The order of strength of intermolecular forces is van der Waals forces < dipole–dipole forces < hydrogen bonding. To recognise the type of intermolecular forces present, remember the three key points:

- All covalent molecules show van der Waals forces.
- To show dipole–dipole forces, there must be elements of differing electronegativity present.
- To show hydrogen bonding, hydrogen must be present and directly attached to nitrogen, oxygen or fluorine.

Types of crystal

A crystal is a solid with particles organised in a regular structure. Consider the following crystal structures:

- **sodium chloride**
- **iodine**
- **diamond**
- **graphite**

(a) Draw simple sketches to illustrate the bonding found in these crystal structures. (8 marks)

(b) State the type of particle found in each type of crystal and the force of attraction that holds these particles together. (8 marks)

(c) Name the crystal structure (sodium chloride, iodine, diamond or graphite) that exhibits each of the following properties and give a brief explanation of this property in terms of structure and bonding:

(i) sublimes on gentle warming

(ii) conducts in the solid state

(iii) very hard, with a very high melting point

(iv) conducts only when molten or in solution (12 marks)

(d) State one other type of crystal structure and give an example. (2 marks)

Total: 30 marks

Grade-A answer to Question 8

(a) The following sketches represent the crystal structures.

Sodium chloride

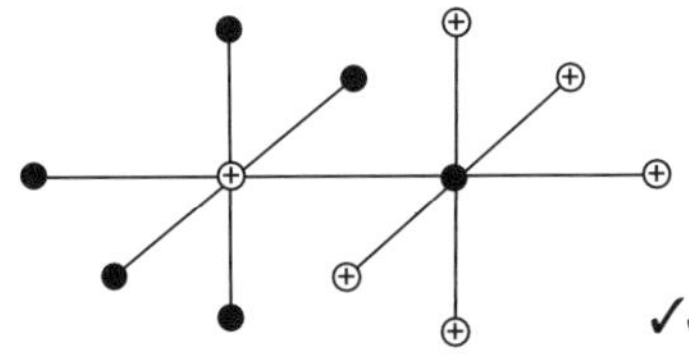

✓✓

Iodine

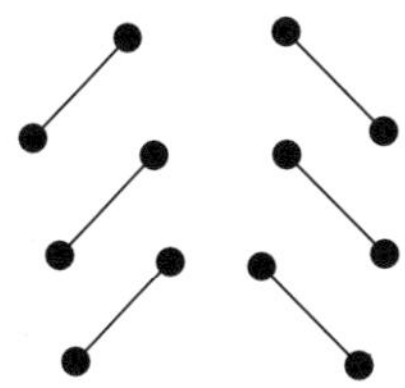

✓✓

Diamond

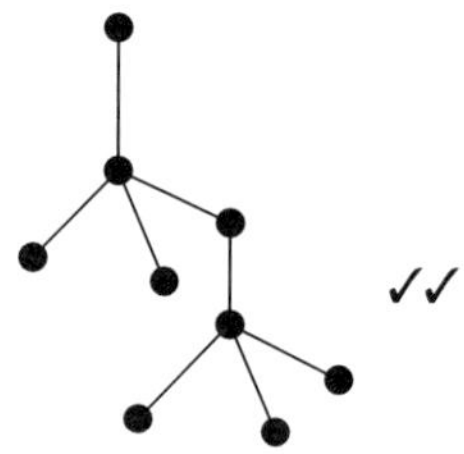

Graphite

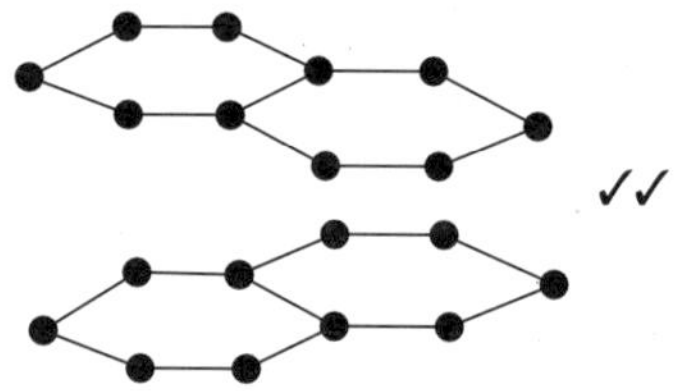

The sketches do not have to be to the same standard as those in textbooks. They need to be good enough to convince the examiner that you understand the key features of each crystal lattice. The sodium chloride lattice needs to show *oppositely charged ions in a lattice* for the first mark and the idea of a *three-dimensional structure* for the second mark. An alternative sketch is shown below.

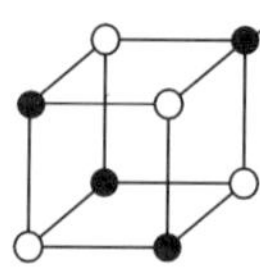

Iodine is a difficult structure to draw, but the simple sketch above is allowed and gains the first mark for the fact that iodine exists as pairs, and a second mark for showing a regular pattern. The structure of diamond must show a carbon attached to four other carbons for the first mark, and at least one more carbon attached to imply that it is macromolecular for the second mark. You must be careful not to draw carbon atoms with more than four bonds. In graphite, the first mark is awarded for implying that it is macromolecular by correctly linking together at least two hexagons, and the second mark is awarded for showing that the structure has different layers, so another hexagon needs to be drawn in a separate layer. A common mistake is to join the hexagons at the corners.

(b) Sodium chloride contains ions ✓, which are held together by strong electrostatic forces of attraction ✓.

Iodine is made up of simple molecules ✓, which are held together by van der Waals forces ✓.

Diamond is a macromolecule made up of many carbon atoms ✓, linked together by strong covalent bonds ✓.

Graphite is another macromolecule made up of many carbon atoms ✓, with strong covalent bonds in the layers and weak van der Waals forces ✓ between the layers.

Diamond and graphite are macromolecules with the type of particle that makes up the crystal being carbon atoms. It is essential that you mention the van der Waals forces between the layers in graphite for the final mark to be awarded.

(c) (i) Iodine ✓ sublimes on gentle heating because to separate the molecules the van der Waals forces ✓ need to be overcome, and these are very weak ✓.

(ii) Graphite ✓ is the only crystal structure listed here that can conduct electricity in the solid state, because some of the electrons ✓ are delocalised ✓ and free to move.

(iii) Diamond ✓ is very hard with a very high melting point because all the carbon atoms are linked by covalent bonds ✓, which are very strong ✓.

(iv) Sodium chloride ✓ only conducts when molten or in solution. This is because the ions ✓ vibrate about a fixed position in the solid state, but when molten or in solution, they become mobile ✓.

Use the question and the mark allocation to structure your answer. There are 12 marks available and four properties to discuss. This means there is 1 mark for the example and 2 marks for the explanation. In the explanations, you must be very careful not to use the wrong words. For example, for part (i) 'Iodine sublimes on gentle heating because to separate the *atoms* the van der Waals forces need to be overcome, and these are very weak' would score no marks because van der Waals forces do not exist between iodine atoms. Another example of a potential slip would be 'Sodium chloride only conducts when molten or in solution. This is because the *atoms* vibrate about a fixed position in the solid state, but when molten or in solution they become mobile'. This would score no marks because sodium chloride is ionic and therefore atoms must never be mentioned.

(d) Another type of crystal is metallic ✓. An example of a metallic lattice is magnesium ✓.

Any named metal will be a suitable answer for the example of a metallic lattice. There are four types of lattice: ionic (sodium chloride), molecular (iodine), macromolecular (diamond or graphite) and metallic (magnesium). If you are ever asked to describe four types of crystal lattice, do not include both diamond and graphite in your answer.

Shapes of molecules and ions

(a) Nitrogen has an atomic number of 7 and a mass number of 14. Consider the following nitrogen species:

- **NH_4^+, ammonium ion**
  NH_3, ammonia
- **NH_2^-, amide ion**
- **NH^{2-}, imide ion**
- **N^{3-}, nitride ion**

(i) State the electron configuration of the nitrogen atom and state the number of unpaired electrons in the atom. (2 marks)

(ii) State the number of protons, neutrons and electrons in a nitride ion. (3 marks)

(iii) Explain how an ammonia molecule could be converted into an ammonium ion and state the type of bond formed in this conversion. (3 marks)

(iv) Explain the differences in the H–N–H bond angles found in the ammonia molecule and the ammonium ion. Include in your answer a sketch, the name of the shape of each species, the bond angles and a reason for any difference. (8 marks)

(v) Use the electron pair repulsion theory to predict the shape of the amide ion. Include in your answer a sketch of the molecule, indicating any bond angles and lone pairs. (3 marks)

(b) Consider the following molecules or ions:

- **BF_3**
- **PCl_5**
- **PCl_3**
- **PF_6^-**
- **PCl_4^+**

For each of the molecules or ions, state its shape and give the values of the bond angle(s) found within the molecule or ion. (11 marks)

Total: 30 marks

Grade-A answer to Question 9

(a) (i) Nitrogen has the electron configuration $1s^2\,2s^2\,2p^3$ ✓. It has three ✓ unpaired electrons in each of the p orbitals, i.e. [↑][↑][↑].

A common mistake is to write 1s2 2s2 2p3 or $1s_2\,2s_2\,2p_3$ and this is not acceptable. Nitrogen has three electrons in the p sub-shell arranged as unpaired electrons in separate orbitals, i.e. $2p_X^1\,2p_Y^1\,2p_Z^1$ and not $2p_X^2\,2p_Y^1$.

(ii) The nitride ion, N^{3-}, has seven protons ✓, seven neutrons ✓ and ten electrons ✓.

The number of protons is equal to the atomic number and the number of neutrons is equal to the difference between the mass number and the atomic number. There is a charge of 3– on the nitride ion. Therefore, three additional electrons are present, making 10 electrons.

(iii) The ammonia molecule contains a lone pair ✓, so it can readily accept an H^+ ion ✓, by reacting with an acid such as HCl, forming a coordinate ✓ or dative covalent bond between the nitrogen atom and the H^+ ion.

e It is essential that you recognise the key features of coordinate bond formation, i.e. the presence of a lone pair on the nitrogen atom in ammonia and the ability to accept an H^+ ion.

(iv) Ammonia has four pairs of electrons, so is based on a tetrahedral structure. However, one of the electron pairs is a lone pair, so the H–N–H bond angle is reduced to 107° because lone pair/bond pair repulsion is greater than bond pair/bond pair repulsion ✓. When ammonia is converted to the ammonium ion, the lone pair becomes a bonding pair (the coordinate bond), so there are now four bonding pairs which repel each other equally ✓, giving bond angles of 109.5°.

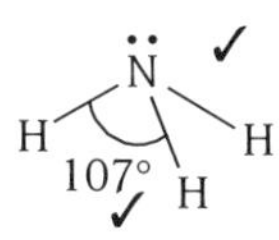

Trigonal pyramidal ✓

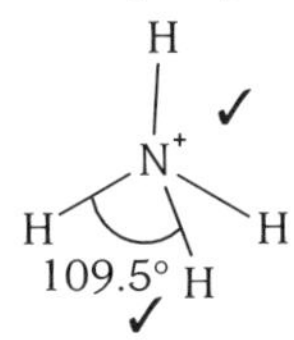

Tetrahedral ✓

e In this question, the mark allocation can be used to determine the structure of your answer. There are two sketches to be drawn, and each needs to be labelled with the name of the shape and a bond angle, giving rise to 6 marks. There are 2 marks left for the explanation and the examiner is looking for your understanding of the key difference between the molecule and the ion, i.e. that both are based on four electron pairs but one has a lone pair present. It is this lone pair which causes the reduction in the bond angle.

(v) The nitrogen has five electrons in the outer shell and it gains two more electrons from the hydrogen in the formation of the amide ion. Add another electron because the ion is negative and this gives a total of eight electrons. Dividing by two gives four electron pairs, so the shape is based on a tetrahedral structure with a bond angle of 109.5°. However, two of the electron pairs are lone pairs, so the bond angle is reduced by about 5°, to give a bent planar molecule with a bond angle of 104.5°.

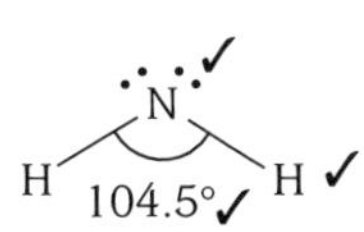

e Amide ions are unfamiliar and many candidates will find this question intimidating. This is an application question, where you need to use the basic principles covered in the section on shapes of molecules. The question asks for a sketch, the bond angle and an indication of the number of lone pairs; these three key points will score 3 marks. There are no marks for the name of the shape or for the reasoning behind the final answer. However, read through the answer again because this explanation is the thought process that you need to go through in order to arrive at the final answer. The answer is very similar to deducing the shape of water molecules.

(b)

Molecule or ion	Shape	Bond angle(s)
BF_3	Trigonal planar ✓	120° ✓
PCl_5	Trigonal bipyramidal ✓	90° ✓ and 120° ✓
PCl_3	Trigonal pyramidal ✓	107° ✓
PF_6^-	Octahedral ✓	90° ✓
PCl_4^+	Tetrahedral ✓	109.5° ✓

e This section only requires the name of the molecular shape and the bond angles. No sketch is required and neither is an explanation of the basic principles. However, this question could easily be expanded to require a sketch or some explanation of how the shapes were deduced. If you score 11 marks on this section, you have a good understanding of the basic principles of shapes of molecules and ions. There are 2 marks available for labelling the two bond angles in PCl_5 — this is why five examples lead to 11 marks. If there is an unexpected additional mark, it is possible that a trigonal bipyramidal structure is one of the answers because this is the only one with two bond angles.

e **Do not be surprised to see questions on other topics when answering a question on shapes of molecules. You will never see a 30-mark question purely on shapes of molecules and ions. It is essential to produce clear, labelled diagrams when drawing simple covalent molecules.**

Question 10

Periodicity

(a) State the electron configurations of sodium and phosphorus and classify each element as an *s*-block or a *p*-block element. State the name and give the full electron configuration of a *d*-block element with an atomic number of less than 30. (5 marks)

(b) Use your knowledge of atomic structure and bonding to explain the following:

(i) the electrical conductivity of sodium is greater than that of phosphorus (3 marks)

(ii) the atomic radius of sodium is greater than that of phosphorus (3 marks)

(iii) the first ionisation energy of phosphorus is greater than that of sulphur (3 marks)

(iv) the melting point of sodium is lower than that of magnesium (3 marks)

(v) the melting point of phosphorus is much lower than that of silicon (4 marks)

(vi) the melting point of phosphorus is lower than that of sulphur (3 marks)

(c) Use your knowledge of the chemistry of period 3 to predict, with a brief explanation, the name of the element in period 2 that has the largest atomic radius, the largest ionisation energy and the highest melting point. A different element should be named for each physical property. (6 marks)

Total: 30 marks

■ ■ ■

Grade-A answer to Question 10

(a) Sodium is $1s^2\,2s^2\,2p^6\,3s^1$ ✓ and is an *s*-block ✓ element. Phosphorus is $1s^2\,2s^2\,2p^6\,3s^2\,3p^3$ ✓ and is a *p*-block ✓ element. Vanadium is a *d*-block element and the electron configuration is $1s^2\,2s^2\,2p^6\,3s^2\,3p^6\,3d^3\,4s^2$ ✓.

e This is a relatively straightforward question and the majority of candidates will score the first 4 marks. Remember to give the full electron configurations. It is essential to use superscripts to show the number of electrons in each sub-shell. In order to score the last mark in this question, you must relate the electron configuration to the stated d-block element. If you state vanadium but give the electron configuration of iron, you will not score this mark. The atomic number must be less than 30, so any correct electron configurations for the elements scandium to copper would be accepted. A common answer will be iron $1s^2\,2s^2\,2p^6\,3s^2\,3p^6\,3d^6\,4s^2$, but two common mistakes occur when writing out the electron configurations of copper and chromium. These two d-block elements are exceptions to the general pattern: copper is $1s^2\,2s^2\,2p^6\,3s^2\,3p^6\,3d^{10}\,4s^1$ and chromium is $1s^2\,2s^2\,2p^6\,3s^2\,3p^6\,3d^5\,4s^1$.

(b) (i) Sodium has metallic ✓ bonding, so because of the mobile electrons ✓, it is a good electrical conductor. Phosphorus (P_4) is a non-metal ✓ and so does not conduct electricity because the electrons are localised ✓ in the covalent bonds of the molecule.

Notice there are four scoring points and a maximum of 3 marks. This is quite common in a question of this type.

(ii) Sodium and phosphorus have the same amount of shielding ✓, but sodium has fewer protons ✓ in the nucleus. This means sodium has a smaller attraction for the outer electrons ✓ and so the atomic radius is larger.

There are three key points to cover. The answer could read 'both have electrons in the same shell, but phosphorus has a higher nuclear charge and so attracts the outer electrons more than sodium'. This covers the same three key points. Examiners are flexible when marking this answer, provided that you can prove that you *understand* these three key points.

(iii) The first ionisations of phosphorus and sulphur both involve removing an electron from the $3p$ level. In phosphorus ($1s^2\,2s^2\,2p^6\,3s^2\,3p_X{}^1\,3p_Y{}^1\,3p_Z{}^1$) the electrons in the $3p$ level are all unpaired ✓, whereas in sulphur ($1s^2\,2s^2\,2p^6\,3s^2\,3p_X{}^2\,3p_Y{}^1\,3p_Z{}^1$) two of the electrons are paired ✓ and suffer repulsion ✓, making it easier to remove an electron.

When answering questions on the drop in ionisation energy between group 5 and 6 elements, always include the electron configuration of both elements. These electron configurations imply that you know the electrons are in the $3p$ level; phosphorus has three unpaired electrons, whereas sulphur has a pair of electrons in one of the $3p$ orbitals.

(iv) Sodium and magnesium are both metals ✓ and so have a lattice of cations held together by their attraction for the delocalised electrons. The metallic bonding is stronger ✓ in magnesium because there is more attraction between the cations and delocalised electrons. There is more attraction because the Mg^{2+} cation has a higher charge ✓ than the Na^+ ion; the Mg^{2+} is smaller (so it has a higher charge density) and it has more electrons per cation holding the lattice together.

This answer contains a lot of detail and 3 marks could be scored for a very simple answer such as 'magnesium has stronger metallic bonding because the Mg^{2+} cations have a higher charge'. The last sentence of the answer contains three reasons why the metallic bond is stronger and any of these would score the final mark. Again, the examiner is testing your understanding of the key points: the type of bonding present, the relative strength of the bonds and the reason for the difference in strength.

(v) Phosphorus is a simple covalent molecule ✓ with only weak van der Waals ✓ forces between the molecules. Silicon is a macromolecule ✓ with strong covalent bonds ✓ between all the atoms.

When explaining melting points, it is essential that you use the word 'weak' or 'strong' when discussing the forces between the particles. 'Phosphorus is a molecule with van der Waals forces and silicon is a macromolecule with covalent bonds' does not answer the question fully. The key point missing is that much more energy is required to break the strong covalent bonds.

(vi) Phosphorus and sulphur are both simple molecular ✓ solids. However, S_8 is a larger ✓ molecule than P_4, so the van der Waals forces between sulphur molecules are much stronger ✓ than between phosphorus molecules.

e When discussing the melting points of simple covalent molecules, it is essential to give the formula of each molecule. The three molecules usually discussed are phosphorus, sulphur and chlorine; the order of melting point is related to the size of the molecules. Argon is made up of single atoms, so it has the lowest melting point in period 3. The order of melting points is $S_8 > P_4 > Cl_2 > Ar$.

(c) The element in period 2 with the largest atomic radius is lithium ✓. Lithium has the same amount of shielding as the other elements, but it has the fewest protons ✓ in its nucleus.

The element in period 2 with the largest first ionisation energy is neon ✓. Neon has the same amount of shielding as the other elements, but it has the greatest number of protons ✓ in its nucleus and so has a greater attraction for the electron being removed.

The element in period 2 with the highest melting point is carbon ✓. Carbon has a macromolecular ✓ structure, with strong covalent bonds between the atoms.

e This question has 3 marks for the names of the elements given for each physical property. If the name is incorrect, then the explanation mark will not be awarded for that particular physical property. Candidates' explanations will vary considerably and sometimes the original marking scheme (pre-exam) will be changed in the post-exam meeting so that common, alternative correct answers are given credit. 'Lithium, because atomic radii decrease across a period; neon, because the first ionisation energy increases across a period; and carbon, because the melting point reaches a peak at group 4' is a simple answer that could score full marks. The important point here is that the candidate fully understands the concept of periodicity and that patterns observed in period 2 are similar to those in period 3.

Group 2 chemistry

(a) Group 2 elements all exhibit metallic bonding.

(i) Describe the bonding in a metal. (2 marks)

(ii) Why is the melting point of calcium higher than that of potassium? (2 marks)

(iii) Why is the melting point of calcium higher than that of barium? (2 marks)

(b) Answer the following questions about the chemistry of magnesium and barium.

(i) State the name of the products formed when magnesium reacts with steam. (2 marks)

(ii) Give the equation for the reaction of barium with cold water. (2 marks)

(iii) Which of the two metals forms the more soluble hydroxide? (1 mark)

(iv) Which of the two metals forms the more soluble sulphate? (1 mark)

(v) State the observations and give equations where appropriate to describe the addition of dilute sulphuric acid to separate solutions of magnesium and barium chloride. (3 marks)

(c) Compare the chemistry of beryllium and other group 2 elements by answering the following questions.

(i) State the type of bonding found in beryllium chloride and barium chloride and account for any difference. (4 marks)

(ii) State the formula of the aqua-cations of beryllium and calcium. State the coordination number of the group 2 element in each aqua-cation and explain any difference. (5 marks)

(iii) Magnesium hydroxide is basic, but beryllium hydroxide is amphoteric. Explain this statement with the use of balanced chemical equations. (6 marks)

Total: 30 marks

Grade-A answer to Question 11

(a) (i) A lattice of cations ✓ held together by a sea of delocalised electrons ✓.

A simple labelled diagram could be used here. Make sure you emphasise the three-dimensional lattice structure of the metal.

(ii) The Ca^{2+} ion has a higher charge ✓ (and a smaller size) than the K^{+} ion, so it has a greater attraction for the delocalised electrons ✓ and therefore stronger metallic bonds.

When an ion has a higher charge and a smaller size, it has a higher charge density. Remember that ionic radii decrease across a period and group 2 cations always have a higher charge density than group 1 cations.

(iii) The Ca^{2+} ion has a smaller size ✓ than the Ba^{2+} ion, so it has a higher charge density and a greater attraction for the delocalised electrons ✓ and therefore stronger metallic bonds.

Ionic radii increase down the group, so Ba^{2+} is larger than Ca^{2+}.

(b) (i) Magnesium oxide ✓ and hydrogen ✓

Magnesium reacts very slowly with cold water to produce the hydroxide and hydrogen, which could be a common incorrect answer. When it reacts with steam, it is the oxide and hydrogen that are produced.

(ii) $Ba + 2H_2O \longrightarrow Ba(OH)_2 + H_2$ ✓✓

Reactivity of the group 2 metals increases down the group. Barium reacts with cold water to produce the hydroxide and hydrogen. There is usually 1 mark for the correct reactants and products and 1 mark for balancing the equation.

(iii) Barium ✓, i.e. $Ba(OH)_2$

Hydroxides become more soluble going down group 2. There is no need to explain the trend.

(iv) Magnesium ✓, i.e. $MgSO_4$

Sulphates become less soluble going down group 2. There is no need to explain the trend.

(v) $MgCl_2$ has no reaction ✓. $BaCl_2$ produces a white precipitate ✓ of barium sulphate:

$$BaCl_2 + H_2SO_4 \longrightarrow BaSO_4 + 2HCl \checkmark$$

Acidified barium chloride or barium nitrate is used as a test for a sulphate — a white precipitate of barium sulphate is formed. An ionic equation is acceptable, for example $Ba^{2+} + SO_4^{2-} \longrightarrow BaSO_4$. The fact that only 3 marks are available for this question suggests that one of the chlorides does not react, i.e. no reaction ✓, observation ✓, equation ✓ equals 3 marks.

(c) (i) $BeCl_2$ is covalent ✓ and $BaCl_2$ is ionic ✓. Be^{2+} is a small cation ✓ with a high charge density, so it is very polarising. It distorts the large Cl^- ion ✓ to such an extent that $BeCl_2$ is considered to be covalent.

The atypical properties of beryllium are caused by its small size and high charge density, which introduces covalency into its compounds.

(ii) $[Ca(H_2O)_6]^{2+}$ ✓, in which Mg has a coordination number of 6 ✓.
$[Be(H_2O)_4]^{2+}$ ✓, in which Be has a coordination number of 4 ✓.
Be^{2+} is a smaller cation than Ca^{2+} ✓, so fewer ligands can fit around the cation.

The coordination number is the number of ligands (e.g. H_2O) surrounding the central cation (e.g. Mg^{2+}) in a complex ion (e.g. $[Mg(H_2O)_6]^{2+}$). All group 2 cations except beryllium have a coordination number of 6. However, beryllium is atypical, with a coordination number of 4, again due to its small size.

(iii) $Mg(OH)_2$ is basic and so dissolves in acids ✓ such as hydrochloric acid:

$$Mg(OH)_2 + 4H_2O + 2H^+ \longrightarrow [Mg(H_2O)_6]^{2+} ✓$$

$Be(OH)_2$ is amphoteric because it dissolves in acids ✓ (hydrochloric acid) and alkalis ✓ (sodium hydroxide):

$$Be(OH)_2 + 2H_2O + 2H^+ \longrightarrow [Be(H_2O)_4]^{2+} ✓$$

$$Be(OH)_2 + 2OH^- \longrightarrow [Be(OH)_4]^{2-} ✓$$

e Many candidates would find this question very difficult, particularly the production of the balanced equations. However, it is still possible to score 3 out of 6 marks for making simple statements regarding the acid–base nature of the hydroxides. The equations are difficult to remember, so you need to think carefully about how they are constructed:

- The hydroxides always have the formula $M(OH)_2$.
- All group 2 hydroxides dissolve in acids to produce the aqua-cation $[M(H_2O)_6]^{2+}$, except beryllium ($[Be(H_2O)_4]^{2+}$).
- Always balance equations for the addition of acids with H^+ and H_2O.
- Only $Be(OH)_2$ dissolves in alkali to produce $[Be(OH)_4]^{2-}$ and only OH^- is needed to balance the equation.

e **The atypical nature of beryllium is a new topic for students and teachers. It comes at the end of Module 1, so the teaching and the revision of the topic may be a little rushed. Many candidates would give poor answers to part (c).**